JAPAN'S SCIENCE EDGE

HOW THE CULT OF ANTI-SCIENCE THOUGHT IN AMERICA LIMITS U.S. SCIENTIFIC AND TECHNOLOGICAL PROGRESS

Shigeru Kimura

February 1985

THE WILSON CENTER
UNIVERSITY PRESS OF AMERICA

Copyright © 1985 by

Shigeru Kimura

University Press of America,® Inc.

4720 Boston Way
Lanham, MD 20706

3 Henrietta Street
London WC2E 8LU England

Co-published by arrangement with
The Woodrow Wilson International Center for Scholars.

The Woodrow Wilson International Center for Scholars subscribes to a
policy of providing equal educational and employment opportunities.

WOODROW WILSON
INTERNATIONAL CENTER FOR SCHOLARS

Smithsonian Institution Building
Washington, DC 20560

Library of Congress Cataloging in Publication Data

Kimura, Shigeru, 1932-
    Japan's science edge.

    Bibliography: p.
    1. Science—United States. 2. Technology—United
States. 3. Science—Japan. 4. Technology—Japan.
5. Technological innovations—United States. 6. Techno-
logical innovations—Japan. I. Title.
Q127.U6K52 1985   509.73   85-5331
ISBN 0-8191-4645-5 (alk. paper)
ISBN 0-8191-4646-3 (pbk. : alk. paper)

CONTENTS

Shigeru Kimura is one of Japan's most distinguished science commentators. A graduate of the University of Tokyo, where he majored in history and the philosophy of science, Mr. Kimura is a recognized expert on nuclear issues. Since joining the Asahi Newspaper Company in 1955, he has held a number of advisory positions and has published over twenty books on scientific and technological matters. In addition to his current responsibilities as the manager of the Analysis and Research Center at the Asahi Newspaper Company, he has been responsible for the Asahi's experiment to make artificial snow crystals under weightless conditions aboard the U.S. space shuttle.

Mr. Kimura completed the major portion of his research for this book when he was a resident Fellow at The Woodrow Wilson International Center for Scholars in Washington, D.C., in 1981.

PREFACE

When I returned to Washington, D.C. in August 1984, I was struck and dismayed by the continuing ambivalence in America concerning the benefits to be derived from investing in America's science and technology base. The "anti-science" disease of the 1960's and 1970's that I examine in this book, symbolized by the "back to nature" ideas of the environmentalists, continues to plague America's technological and economic competitiveness.

Beginning in the youth "counter-culture" protests of the 1960's, "anti-science ideas" were more formalized in 1970 when Nixon administration officials focused national attention on water and air pollution and industrial waste in an effort to shift public concern away from the Vietnam War and civil rights issues. The energy crisis of 1973 and the resulting rise in energy prices in the late 1970's kept attention on these same environmental problems and on the regulation of industry, much to the detriment of America's industrial competitiveness.

These issues were particularly real in 1981 when, as a Fellow at The Woodrow Wilson International Center for Scholars in Washington, D.C., I began my research into the historical and cultural sources of America's technology decline. Earlier in 1980, I had sensed Americans were approaching a turning point regarding their previous doubts about scientific solutions to toxic waste disposal, chemical additives in food, consumer product safety, and other key problems. Balance was being restored in the national objectives: equilibrium between quality of life issues and the economic necessities for survival. People were talking about industrial reorganization, competitiveness, and restoring research and developmental incentives to American industry. My earlier fears that the anti-scientific attitudes and prejudices that flourished in the 1960's and 1970's had contributed to America's economic recession would continue seemed misplaced.

The germs for positive change are still there, but I am not convinced the United States has regained its former faith in the power of science and technology. The

Reagan administration's emphasis on large technical projects like the space stations, and its reliance on the private sector to develop new technologies to improve the economy are having an effect, but it is not enough. However, progress toward improving the public's awareness of science and the new technologies is evident. Although opinion polls show that people again look toward science for solutions to many problems, the media still dramatizes scientific dangers, oversimplifying the issues and exaggerating the risks.

Americans must be made to realize certain realities: the Russians will pursue science in line with their aims of world dominance; and the Japanese will march forward full steam to enhance their economic competitiveness. Ambivalence will not help America to survive in the face of such a rigorous two-dimensional competition. Legitimate environmental concerns that restrict the use of new technologies must be weighed against the costs that they impose on the economy. Preeminent national strength is impossible to sustain in a capitalist system without a vigorous economy and the technological progress needed to be competitive in the world market.

Philosophically, I support scientific and technical progress. Like most Japanese, I also hope for a stronger and more prosperous America. With this book I have tried to diagnose the ailment, isolate the virus, and suggest some remedies. As a science writer for the leading Japanese newspaper, the *Asahi Shimbun*, I believe in the appropriate and adequate attention by our two countries to the scientific problems at hand. We all appreciate that we must be cautious about the abuses of science, but we should not lose sight of the overall benefits. In this book, Japan's experience with a shorter bout of the same anti-science illness is compared to the case I examine here for the United States. Japan is now doing far better than the United States in balancing the costs and benefits of science and technology. That is why Japan is so economically competitive. Americans must make the same hard choices.

Being able to share these thoughts would never have been possible were it not for the kindness and support of the staff of The Wilson Center. Prosser Gifford, Deputy Director of the Center, and Ronald Morse, Secretary of the Asia Program, assisted in more ways than I can mention. Of special help in Washington were Dr. Edward J. Burger, Jr., a former White House science adviser, and John M. Logsdon, Director, Graduate Program in Science, Technology, and Public Policy at George Washington University. Research assistants, provided by The Wilson Center, were far more helpful than they realize. I would like to thank Anne Harriman, Timmie Rony, and Kanako Yamashita for their hours of dedicated assistance. Kathy A. Meyer of The Wilson Center patiently typed the manuscript and Paul Gebhard and Helen Loerke assisted in the editing. To the many other colleagues and friends who made my stay in America enjoyable, I express my sincere thanks.

Shigeru Kimura
August 1984

I
AMERICA IN TROUBLE

In 1984 the launchings of the space shuttle barely attracted any popular attention. Owners of local restaurants and tourist concessions at the launch site complained about the lack of business. This contrasts markedly with my earlier experiences in 1981 and 1969. In April 1981, a million spectators from all over the United States surrounded the Kennedy Space Center, anxious to catch a glimpse of the launching of the space shuttle *Columbia*. Hotel and motel rooms that normally rented for as little as $20 per night commanded $50 or more for that one night. Signs indicating "No Rooms Available" were displayed everywhere.*

All along the sweeping shoreline, campers and automobiles were parked side by side. Police handed out warning notices to the arriving spectators: "Be sure you have a full tank of gas and enough food and water. Once your vehicle has entered the traffic jam, it will not be able to leave until after the launching." By 1:00 A.M., the spectators' cars, vans, and trucks had completely clogged the roads, making it impossible to move more than one mile an hour. The time it took to go from our motel to the press corps seats inside the Space Center was normally only 20 minutes but now it took us more than four hours. Many of the reporters and photographers were worried about reaching the press area in time for the launching.

Removed from this confusion, inside the Space Center preparations for the launching of the space shuttle had been moving steadily ahead. Because of computer trouble, the launch had been postponed for two days, until 7:00 A.M., April 12, 1981. Four thousand five hundred writers and photographers from every part of the world gathered to keep a vigil throughout the night. Despite what one had heard about "tropical" Florida, the temperature by dawn was uncomfortably cool. The space shuttle itself stood on Launching Pad 39A, about 3.5 miles away from the press boxes. From this same pad, on July 16, 1969, the spacecraft *Apollo 11* had blasted off toward the moon. Like the Saturn 5 rocket that had lifted *Apollo* into space, the shuttle resembled a giant white whale. As the sky in the east became lighter, a feeling of tension increased among the journalists. The huge digital clock mounted in front of the press stands ticked away the minutes and seconds.

"Five minutes until liftoff . . . four minutes until liftoff . . ." Then, at last, the final minute. "Ten seconds to go, five seconds, four, three, two . . ."

*All references are in the Sources section at the end of the book.

The space shuttle spouted a blinding orange flame, and a tremendous billow of white smoke concealed, for an instant, the huge shape of the spacecraft. It began slowly to lift into sight. The journalists rose from their seats clapping and cheering. The immense sound of the rockets shook the roof shingles of the press stands, and the earth trembled violently.

The space shuttle, atop a white trail, rose gently into the clear morning sky. Television monitors in front of the press stands showed the booster rockets falling away from *Columbia*. The members of the press seized their telephones and bent over their typewriters. The scene reminded me of my experience fifteen years earlier at the launching of *Apollo 11*. Looking back, *Apollo 11* now seemed a flower of America's golden age.

Apollo 11's lunar module, which was left on the moon's lonely surface, has a plaque engraved with the words: "We came in peace for all mankind." Commander Neil Armstrong and his two fellow astronauts really did go to the moon as representatives of mankind. On that day I sent the following dispatch to my newspaper, the *Asahi Shimbun* in Tokyo:

> The more than 3,000 reporters and photographers who had assembled on the grandstand at the Kennedy Space Center were no longer simply journalists. As the *Apollo* spacecraft lifted off from the launch pad and headed for man's first landing on the moon, they whistled, cheered, and clapped furiously, shouting, "Go! Go!" The cool, professional objectivity of these journalists had suddenly melted away. The reporters who leaped from their seats and shouted "Go! Go!" were not only Americans, but Swiss, Frenchmen, Africans — everyone present was carried away by wild enthusiasm. They were simply a crowd of human beings, giving a sendoff to three fellow human beings.

At that time, fifteen years ago, American science and technology outdistanced that of Japan or any European country. American science and technology was in every respect the finest in the world. America was truly a giant, the representative of mankind. But during the fifteen years since then, America seems to have greatly weakened, to have lost her vitality. Americans themselves have lost their self-confidence.

When the space shuttle *Columbia* left the launching pad, President Reagan lay in a hospital in Washington. Two weeks earlier, he had been wounded in the chest by a 25-year-old youth. The President, however, sent a message from his sickbed to the two astronauts aboard *Columbia*:

> Through you, today, we all feel as giants once again. Once again we feel the surge of pride that comes from knowing we are the first and we are the best and we are so because we are free . . . . As you hurtle from earth in a craft unlike any other ever constructed, you will do so in a feat of American technology and American will.

Why did President Reagan feel compelled to twice repeat the words "once more" in this message? Why was there no need before now to emphasize that America is a "giant"? The reason, perhaps, was that in many areas of technological development, America had already lost her position as "Number One." Dr. Simon Ramo, of TRW, Inc., one of America's advanced technology companies, published *America's Technology Slip* in 1980. There he wrote:

> But today we no longer can assume we are ahead. Contrary indications are all about us in the form of European and Japanese cars on our streets and foreign-made television sets and tape recorders in our homes. We are lagging badly in other fields and being overtaken in some areas where we still have a lead.

In 1981 abundant evidence of this could be seen at the Kennedy Space Center. To be sure, no other country but the United States would have been capable of building the space shuttle. It is the world's finest spacecraft, and America has every right to be proud of it. Yet virtually all of the thousand or more press cameras that photographed *Columbia*'s ascent were made in Japan. Nearly all of the video cameras carried by television cameramen were made in Japan. One television newsman told me: "Japanese TV cameras are the best in the world. We have to wait six months after we order them, but they're worth it, because they really are the best." It sometimes seems almost as if Japanese technology has been gradually improving while American technology has actually deteriorated. The veteran science reporter of the *New York Times*, John Noble Wilford, wrote an article called "Space and the American View," published about one week before the launching of the space shuttle. He commented:

> Rather, it is that our technology does not seem to work as well as we used to think. Automobiles are always being recalled because of defects; cities are blacked out by power failures; arena roofs collapse; design flaws ground a fleet of jetliners; a nuclear plant breakdown causes near panic; even the helicopters for the attempted hostage rescue mission fail. And buses made by the same manufacturer that built the lunar module — that genuine model of zero-defect engineering — cannot stand up to a few weeks on New York City's streets. Have we lost, we ask, the technological touch?

Of course, the technological lag is not the sole factor in the American people's loss of self-confidence. As Wilford also points out, America was forced to pull out of Vietnam, and later was powerless for more than a year to resolve the Iranian hostage incident. America has been unable to prevent the OPEC countries from raising the price of oil virtually at will. Various episodes of this kind have helped to drag America from her position at the very top. But not the least of the causes, in my opinion, is America's scientific and technological decline, whose significance cannot be underestimated. An American journalist in the press section at the Space Center put it very aptly when he said to me: "President Reagan's feeling that we've become 'giants once again' is only going to last for the two and a half days the space shuttle is in orbit. This country may never be a giant again." From what I have seen since 1981, it is very possible that America does not perceive the combination of policies really needed to accomplish the greatness it was once thought capable of. As we look to 1985 and beyond, the prospect for rejuvenation in America is not good.

Japan Surges Ahead

When one compares the technologies of the United States and Japan, America's relative "slippage" becomes obvious. In June 1981, Lionel H. Olmer, the U.S. Under Secretary of Commerce for International Trade, said the following about trade with Japan. His comments in 1984 were not much different:

> A number of people have observed that even at present we have become a developing country with respect to the Japanese. We ship them raw materials, and they send us manufactured goods.
>   Our imports from Japan are now 50 percent larger than our exports to Japan. We find ourselves having to run faster just to prevent back-sliding. Since 1970, our exports to and imports from Japan have each been growing at about 17 percent a year. If this trend continues, our deficit with Japan will be $22 billion in 1985 and $48 billion in 1990.

After strongly criticizing Japan for *de facto* import restrictions on oranges and communications equipment, Olmer went on to say: "The Japanese are in every area that is the wave of the future. They are not dominant in the leading edge of semiconductors and computers, but they are in those areas, and it would be absolutely foolish to underestimate their competence."

Moreover, said Olmer, he has asked many people the following question: "Would it be a matter of concern if by the end of this decade the U.S. Defense Department had to rely on imports from Japan for substantial quantities of computers, semiconductors and telecommunication devices?" Olmer himself believes that the possibility of this happening is at least 50 percent.

The steel industry in the United States today is about to disappear. In the latter half of the 1960's, Japanese steel manufacturers caught up with and passed their American competitors. The Japanese firms had made an effort to modernize their factories by introducing computers and automatic controls in many steelmaking processes. In addition, they had attempted to reduce their use of energy, whether from oil, coal, gas, or electricity, in making steel.

What was the result of these efforts? Dr. Simon Ramo has written: "In the mid-1960's it took some 25 worker-hours to produce a ton of steel in a Japanese mill while the labor required in the United States was half that amount." The unit "worker-hour" (or "man-hour") refers to an amount of labor: that is, the amount of work performed by one worker in one hour. Thus, in the middle years of the 1960's, only 12.5 man-hours of labor were needed, on the average, to make a ton of steel in an American mill. But then, Ramo continues:

> By the mid-1970's the U.S. figure improved to 10 worker-hours, but the Japanese meanwhile brought theirs down to 9. Japanese steelmakers

have facilities much more efficient than ours — big, well laid out plants using the latest technology, large blast furnaces, more extensive use of continuous casting, better integrated operation, and expanded computer control. In the past 20 years, Japan has built several completely modern plants; the United States has built one. The energy consumed to make one ton of steel in Japan is now about a third lower than that in the United States.

Ramo's description becomes even more striking in graphic form. Figure 1 compares worker productivity in the steel industries of several countries over a period of 18 years. The graph shows how many tons of crude steel a single worker in a given country could produce in one year. The figure in parenthesis next to the name of each country represents the factor of increase in worker productivity for that country's steel industry during the years 1965-83.

Japan's steel productivity, it can be seen, rose at a factor of 3.2, while that of the United States increased only 1.3. As the graph shows, Japan's productivity overtook that of America in 1967, and the gap continued to widen thereafter. The amount of energy needed to manufacture a ton of steel in Japan, the United States, and three other countries is shown in Figure 2. Total energy consumption, indicating the consumption in Japan in 1973 per ton of steel produced, is shown as 100. In this graph, the smaller the number, the better the energy-use technology of the country's steel industry. If Japan's energy consumption in 1981 is expressed as 100, then America's consumption in the same year is 148. Thus, as Ramo indicated, America's steel industry uses about 50 percent more energy than Japan's.

Any increase in the cost of energy sources like oil or coal is quickly reflected in steel prices. Figure 2 suggests that the price of American steel is more easily affected by oil price increases than is the price of Japanese steel.

Today, Japanese steelmakers are not so concerned about higher oil prices. Says a Japanese steel technologist: "Since the first 'oil crisis' in 1973, we have been pouring all our efforts into energy conservation. As a result, we are now able to make the same amount of steel with less energy than any other country. It may sound paradoxical, but the higher oil prices go, the cheaper Japanese steel becomes, compared to steel made anywhere else."

These Japanese scientists and engineers have developed many new techniques for saving energy, and have introduced them at the factory level. When these Japanese achievements are described, many Americans tend to say something like this: "But Japan made a fresh start after the second World War. The Japanese got started later than America, so they were able to introduce newer types of equipment." The second World War, however, ended in 1945. Japan did make an effort to bring in the newest technology from abroad, but as Ramo indicates, the Japanese steel industry's productivity until the first half of the 1960's was less than one-half that of the United States. Japan's productivity surge from that point on was totally unrelated to her so-called "later start." Ramo goes on to ask:

Is it the lack of enough management skill, R & D, and technological innovation that is causing our steel industry to drop in world stature?

6

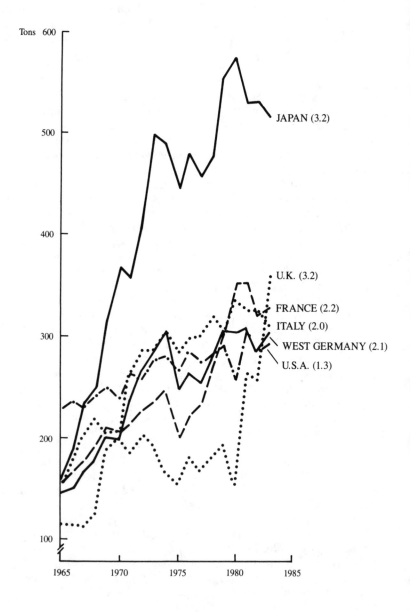

Figure 1   ANNUAL STEEL PRODUCTION PER WORKER

7

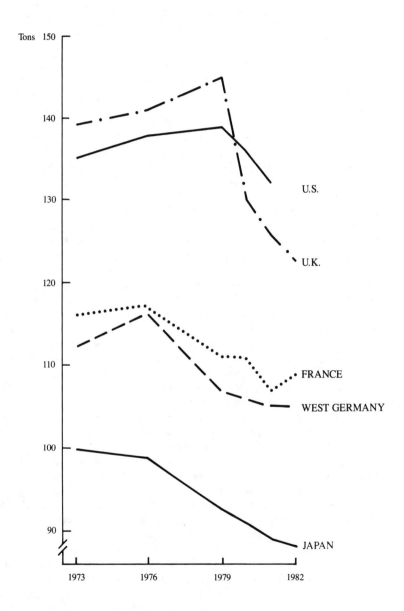

Figure 2   ENERGY CONSUMED IN STEEL PRODUCTION

Not anymore, because any valuable process or technique for manufacturing steel used in Japan and Germany could be incorporated here as well. American steel companies are not now choosing to introduce certain advanced technological methods because of anticipated low return on the incremental investment and an inability to raise capital in view of the poor returns.

In the field of numerically controlled machine tools, Japan overtook the United States in 1979. Machine tools are used to cut, shave and bend such materials as steel or aluminum plates and sheets, in order to make all kinds of products and parts. Suppose, for example, that a doughnut-shaped part has to be cut from a large steel plate. Formerly, concentric circles would be drawn in ink on the surface of the plate, and a skilled operator would carefully guide the metal-cutting saw blade of a machine tool along the circles of ink. Because the saw blade was controlled by a human hand, the operator's level of skill would determine whether the resulting doughnut-shaped plate was correctly and precisely cut.

With the development of the computer, however, machine tools have changed completely. Computers are now able to control machine tools automatically. It is no longer necessary, for example, to draw concentric circles in ink on a steel plate. Shapes are stored in a computer as a series of numbers, and the machine tool moves as instructed by those numbers. Since the numbers control the machine, this technique is known as numerical control, or "NC."

Figure 3 compares Japanese and American statistics on machine tools. The upper graph shows total annual deliveries of NC machine tools. In 1975, only one-fifth as many NC machine tools were delivered in Japan as in the United States, but by 1979, Japan had pulled very slightly ahead of America. The lower graph shows what percentage of the total number of machine tools delivered were equipped with numerical controls. Until 1977, the ratio was higher in the United States. In 1978, however, Japan's ratio rose higher than that of the United States.

In the use of industrial robots, as well, Japan is ahead of any other country. These robots, bearing no resemblance to human beings, look more like a rectangular box with a single arm attached. With this arm a robot might perform welding, tighten screws, or move parts. These robots are tremendously hard workers. They require no coffee breaks or holidays, and can maintain their silent, steady pace 24 hours a day, seven days a week. Robots are never confused by model changes in the cars they build, for example, because their work routine can easily be reprogrammed to suit the new design. Japan imported her first industrial robots from the United States in 1968. She then began domestic production of robots, and now has around 70,000 in use, with their numbers increasing by about 20,000 each year. Eighty percent of all robots at work in the world today are in Japanese factories.

One reason for the rapid increase in the number of industrial robots in Japan is the labor situation there. With a declining birth rate, it is harder to find younger workers; and as the average educational level of the Japanese people has continued to rise (the ratio of students going to senior high school — grades ten through twelve — is expected to reach 99 percent by 1985), the number of blue-collar workers has been declining. Yet robots have not been introduced simply to compensate for a labor shortage. Unlike human workers, robots never tire, so their work is uniform and the

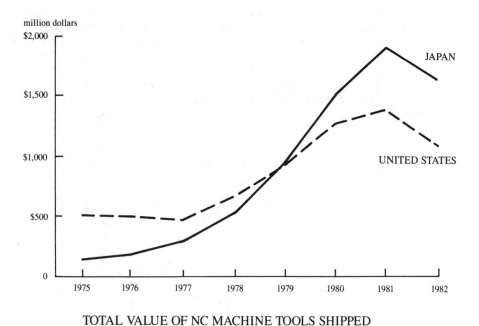

TOTAL VALUE OF NC MACHINE TOOLS SHIPPED

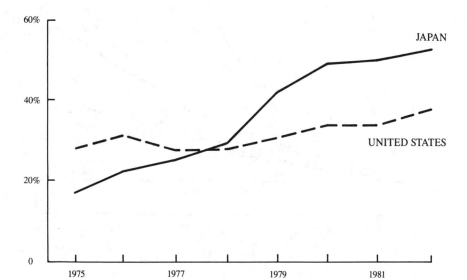

PERCENTAGE OF NC MACHINE TOOLS AMONG ALL MACHINE TOOLS SHIPPED

Figure 3

products they manufacture are higher in quality. Also, they are less expensive than human workers from a wage standpoint, so they tend to reduce the cost of finished products. Thanks in great part to automation of this kind, Japanese labor productivity has been rising sharply for the past several years. This increase is shown in Figure 4. In the period 1960-80, Japan's productivity multiplied nearly fivefold. By contrast, America's productivity merely doubled.

America's technological decline can be seen most clearly when one watches the automobiles that crowd her streets and highways. When *Apollo 11* landed on the moon in July, 1969, hardly any Japanese-made cars were to be seen on America's

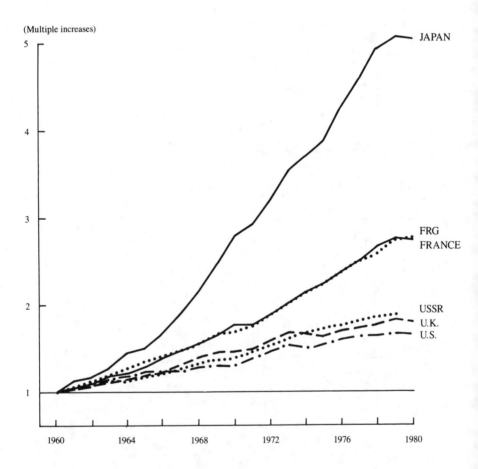

Figure 4   LABOR PRODUCTIVITY IN MANUFACTURING

roads. Most were full-size, heavy American cars like Chevrolet and Cadillac. One occasionally noted a Volkswagen, but such compact and sub-compact cars were no more than a minority group. But in 1981, when *Columbia* was launched, more than one-half of the cars on the road in America had become compact-size models. Of these, the most popular were Toyotas, Datsuns, and other Japanese cars. Right around the time of *Columbia*'s initial voyage, the newspapers were full of stories and articles about the issue of restricting exports of Japanese cars to the United States. These export restrictions are still in effect today. In 1980, the total number of Japanese cars brought into the United States had climbed to 1,820,000 units. Sales of American-made cars, hard-hit by the Japanese imports, had slumped, and the Big Three automakers had shown a total loss, for 1980 alone, of $4 billion.

Automobiles, like steel, were one of America's key basic industries. At one time, 75 percent of all automobiles made in the entire world were manufactured in the United States. Yet, for the year 1980, the number of cars built in Japan actually exceeded the number built in America (see Figure 5).

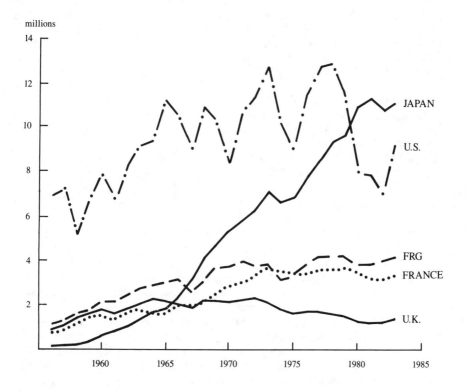

Figure 5   PRODUCTION OF AUTOMOBILES

For American car makers, the greatest shock was how quickly and easily the preferences of American buyers changed. Because of the sudden increase in gasoline prices following the first oil crisis in 1973, the American public's tastes moved away from full-size cars that consumed large quantities of gasoline, toward smaller cars that could travel farther on less gasoline. Japanese auto manufacturers had always concentrated on small cars. They also responded to the 1973 rise in oil prices more quickly. Because Japan is a small, crowded country with many narrow streets and roads, compact-size cars are better suited to Japanese conditions than larger ones. Thus, Japanese small cars had become highly reliable and refined technologically.

American car makers had been unable to build good small cars in response to consumer demand since 1973. The small cars hastily designed and manufactured by Detroit tended to be mechanically unreliable. Many American buyers seemed to be saying: "Japanese cars are much higher in quality, and they break down less often. The best thing is that they cost a lot less to repair." At one time, the label "Made In Japan" symbolized cheap, low quality, poorly-made products. For a while after the end of the second World War, the goods exported by Japan to the United States consisted primarily of flimsy toys and low priced miscellaneous merchandise. Today, however, the high quality of products "Made In Japan" is recognized by the American consumer. The New York Times, on May 10, 1981, carried an article entitled, "He Taught the Japanese." "He" is Dr. W. Edwards Deming, who is probably more familiar to virtually everyone in the Japanese industrial world than he is in the United States. Deming can truly be called Japan's "god of quality control." Japanese industries invited Deming to Japan in 1950 to teach his technique of quality control. His method was the very latest one, utilizing statistical analysis. Soon thereafter, the "Deming Award" was established in Japan, and is still presented annually to the firm with the most successful quality control programs. In the 30 years of the Deming Award's existence, more than 100 Japanese companies have received this honor. The New York Times reports that "The honor roll includes such major corporations as Nissan Motor, Toyota Motor, Hitachi, Matsushita Electric, and Nippon Steel."

Asked on January 15, 1984 (The Washington Post) if anyone from the White House had even called him, Deming replied, "No, why should they?" Responding to the point that those people are concerned about U.S. competitiveness, he said, "So concerned that they don't know anything about what's going on? . . . So concerned that they have no idea about the transformation that must take place?" Still, Deming is in great demand in the United States, and is invited to seminars and symposiums all over the country, where he discusses quality control techniques with American managers anxious to use these "old American techniques." The New York Times notes: "American executives, lulled by the era of easy economic growth that lasted into the early 1970's, have often ignored home-grown methods like Mr. Deming's quality control techniques. Now, however, they are scrambling to re-learn some neglected skills." Deming is quoted as saying, "American industry can do a whole lot better than the Japanese if we get going." But in the area of quality control, as well, America can probably be considered to be at least ten years behind Japan. The New York Times reports incidents like the following:

Robert Quarti of Kensington, Conn., decided to take advantage of a
$500 rebate offer last month to buy a new Chevrolet Citation, one of

the "X" series of cars sold by the General Motors Corporation. As he drove home from work a few days later he noticed a rattle in the dash panel. When he investigated, the whole panel slumped toward him. "I took it to the dealer and he told me the bolts that hold it on had never been tightened at the factory," Mr. Quarti said. "It's sort of disappointing to have this kind of thing happen when you pay almost $8,000 for a car."

I sighed when I read this article, for I, too, have had a similar experience. I had arrived in Washington in February, 1981, and rented a furnished apartment. It consisted of a single room in a nine-story building, and contained an impressive sofa, a stylish table, and a large bed. I kept waking up in the middle of the night, however, because whenever I turned over in my sleep, the bed would emit a loud creaking sound. When I investigated, I discovered that the bed was of the folding type, and that the bolts that locked the segments together were loose. It was not a matter of the bolts having worked loose after having been tightened — no, it was obvious that they had never been tightened at all. The table also wobbled shakily, spilling my coffee. The bolts holding it together were loose, too. I had to go out and buy a screwdriver and a wrench to tighten all of these bolts. Failure to tighten bolts is a problem that existed long before quality control, and comes under the heading of negligence. As long as it is limited to such items as beds and dashboards, this kind of carelessness might be dismissed as only a minor offense. But with many products, the consequences can be much more serious.

In November, 1979, for example, at the Takahama nuclear power plant in Fukui Prefecture, Japan, a large quantity of water leaked from the reactor and caused a major uproar. An inquiry revealed that the water had leaked from a hole that had been cut in a pipe in preparation for installing a flow meter. It turned out that the meter would not be needed for the time being, so the hole had been covered with a plug. The plug cracked, however, allowing water to spurt from the pipe. Plugs of this type are supposed to be made of stainless steel, but the actual plug used in this case, imported from the United States, was made of a copper alloy — even though it was stamped "stainless steel." Copper alloy does not withstand heat and pressure very well, so the plug soon broke. Since that incident, every single American-made part or component in any Japanese nuclear power facility has been re-inspected.

The 1984 presidential election focused on industrial policy issues — issues that have gone neglected for the last three years. Looking again at automobiles, the *New York Times* and CBS News conducted in the spring of 1981 a public opinion poll dealing with cars. Among the questions asked was this: "Do you think that Japanese-made cars are usually better quality than those made here, about the same quality, or not as good?" Of those asked, 34 percent said "better quality," while 30 percent answered "the same." "Not as good" was the reply of 22 percent, and the remaining 14 percent did not know. When a Roper Poll had asked the same question in 1977, only 18 percent answered "better quality." "The same" was the reply of 30 percent, "not as good" of 32 percent, and 20 percent did not know.

These figures indicate a rapid improvement in the reputation of Japanese cars over three or four years. Thanks to this enhanced reputation, the Japanese automakers no longer need to go into detail about the features and quality of their cars when they advertise in newspapers, magazines, and television. From the spring of 1981 the Japanese companies appeared to be selling their cars' "image" in their advertisements and commercials. Now they have moved in to challenge the U.S. luxury car market.

One Datsun television commercial, for example, shows a smart-looking car speeding along a highway, with a background chorus of "We are driven!" Toyota's commercial sings "Oh, what a feeling!" while Mazda uses the slogan "Just one look!" By contrast, the advertising of the Detroit car manufacturers tries to persuade the consumer that American cars are a quality product. The commercials provide A-to-Z laundry lists of items included as standard features, along with specifications like gas mileage. Thus, the previous perceptions of American and Japanese quality have been, in effect, totally reversed.

This American lag in technological development can create social problems. In America's Motor City, Detroit, around 200,000 workers have suffered the hardship of being laid off. The *New York Times*, on April 27, 1981, used the case of one of these workers, Steve Ristoski, to depict the ordeal of the unemployed. Ristoski is 42 years old, with a wife, Angelina, and two sons. Employed in a Chrysler plant, he was put on "indefinite layoff" on October 8, 1979. His wife, who worked for a small parts manufacturer, was then laid off in the spring of 1980. Ristoski, like other workers whose "temporary" layoffs stretch on for months and years, might well have been better off looking for work in another part of the country. But Ristoski hoped that this time Chrysler would call him back to work, as in the past, when the economy improved, so he did not leave Detroit in search of work. He told a reporter from the *New York Times*: "I work seven and a half years in one place, never late, never a complaint. The boss wants overtime, I work it — 10, 12 hours. I don't care, I like to work. I thought Chrysler would call me back." In fact, when Ristoski was laid off once before, during the automobile sales slump in 1974-75, he was indeed called back to work after several months. This time too, he thought, he would only have to be patient for a few months. But all of his hopes have been dashed. "Sure," Ristoski told the reporter, "I would move if I could find a job, but with what? You want to sell your home to go someplace else, you can't sell it because nobody's got a job, nobody's got the money. With no money for gasoline, I can't even leave the house to look for work."

In the Detroit area and elsewhere in America, 200,000 other laid-off auto workers share Ristoski's predicament. The U.S. government tried to get Japan to place voluntary restrictions on her automobile exports to America. Japan finally agreed to limit exports during 1981 to 1,680,000 cars. That represented a 7.7 percent decrease from the 1,820,000 cars shipped in 1980. But what percentage of those 200,000 unemployed auto workers will be put back to work by a 7.7 percent cut in Japanese imports? The technological lag is having a major effect upon American society, and the ordeal of Detroit is a conspicuous example.

Third Wave Technology

Alvin Toffler's book *The Third Wave*, published in 1980, became a sudden worldwide best-seller. It was translated into Japanese that same year, and was widely read in Japan. In his book, Toffler divides the development of human society into three stages. The first, agricultural society, began around 8,000 B.C., and flourished for nearly 10,000 years. From 1650 to 1750, industrial society, called the "second wave" by Toffler, began to replace agricultural society. Finally, from about 1955, a new "third society" appeared. According to Toffler, typical products of the "second wave" include coal, railways, textiles, steel, automobiles, rubber, and machine tools. In the "third society," says Toffler, the important roles will be played by four technological groups: computers, space industry, ocean development, and genetic industry. I cannot agree with everything Toffler says in his book, but if one assumes that his theories are correct, then the industrial sectors I have already cited in which America has been overtaken by Japan — steel, machine tools, automobiles — are technologies of the "second society."

Then is America a leader in the technologies of what Toffler calls the "third society"? In the space industry, certainly, America stands foremost in the world. Many predict that the space shuttle is already bringing immeasurably great benefits to industry. This is because of the state of weightlessness that exists within the space shuttle, and because of the ultra-high vacuum of space itself, which is freely available from the spacecraft. Some American scientists say that from now into the twenty-first century, 80 percent of all major inventions and discoveries will be made aboard space shuttles.

America's National Aeronautics and Space Administration (NASA) very much welcomes participation by other countries in experimentation aboard the space shuttle. For a fee, NASA has offered to carry equipment for experiments on board the shuttle, but Japanese firms have not yet indicated an enthusiastic response. However, once the space shuttle is on a regular schedule, making flights, say, once each week, then NASA may find Japanese firms beating down its doors. As for ocean development, President Johnson called for the application of space technology to the world's oceans, and in the mid-1960's great strides were made in this area. But little progress has occurred since then. Genetic industry is still in its infancy, and it is hard to judge which countries are ahead or lagging behind. In any event, neither space industry nor ocean development nor genetic industry has yet developed to the point of being commercially viable on any extensive scale.

Of the four industries of the future listed by Toffler in his book, the only commercial success so far is the computer industry. Many people are already familiar, from their daily lives, with the rapid developments in computers. Seats on airplanes and trains can be instantly reserved by computer, while money deposited in a bank can be withdrawn simply by inserting a card into a slot and pushing a few buttons. In the field of computers, America's IBM Corporation has dominated the world for many years. IBM, however, is being pursued by Japanese companies, and is gradually being overtaken. Semiconductors, the essential components of all of today's computers, are another area in which America is hard-pressed by Japanese industry. The semiconductors most commonly used in computers are the so-called 16K RAM (16 kilobyte Random-Access Memory) elements. These have layers of complex circuits on a square chip of silicon no larger than the nail of one's little finger.

The term "byte" denotes a unit of information. One page of this book, for example, contains about 10,000 "bytes" of information. Sixteen "kilobytes" (16K) means 16,000 "bytes" that one 16K RAM chip can store — an amount of information equivalent to approximately two pages of this book. Now technology is moving into whole new levels of memory storage.

The total value of all semiconductor elements produced in 1980 was about 8 billion dollars, with 40 percent of world sales going to Japanese manufacturers. The successor to the 16K RAM chip is the 64K RAM chip. This semiconductor element can hold 64,000 "bytes" of information, corresponding to approximately eight pages of this book. Japan, it appears, will be able to capture a major share of the world market for this new product as well. Next down the road is the 128K chip, which will contain information equal to more than 16 pages of this book on an element the size of one's smallest fingernail. As chips of ever-increasing storage capacity have become available, computers themselves have grown smaller in physical size, while their capabilities have simultaneously increased. This process can be expected to continue. Small computers will become convenient for use in the average home. As this happens, the demand for semiconductor elements will grow. American specialists predict that the value of total semiconductor production in 1990 will reach $55 billion. Even in this supergrowth industry, however, American semiconductor manufacturers have been forced into a painful predicament since the latter half of 1980. For example, the profits of Texas Instruments, Inc. for the first quarter of 1981 were 32 percent lower than in the same quarter of 1980. On May 19, 1981, the *New York Times* published a graph comparing the total value of exports of semiconductors, in each direction, between the United States and Japan, from 1975 to 1980. Based on data from the U.S. Department of Commerce, the graph is reprinted here as Figure 6, and shows how Japan overtook America in 1977.

Figure 6   U.S. AND JAPANESE SEMICONDUCTOR EXPORTS

How did this happen? Representatives of American semiconductor manufacturers provided a fairly detailed explanation to some reporters and editors of the *New York Times*. The *first* reason is superior techniques of Japanese companies. Some of the semiconductors manufactured by any factory are bound to be rejected, but the reject rate from Japan's best factories was only three units out of every one thousand made, while in America ten out of one thousand — or more than three times as many — were rejected by customers. Because of the high cost of the silicon from which the chips are made, the rejection rate affects the price of the finished semiconductors. One American chip manufacturer's representative interviewed by the *Times*, however, said of the American problems with rejects: "That's two years ago, that's a dead issue. . . .Today, any customer you talk to will tell you there's no difference between the best Japanese supplier and the best American supplier."

A *second* reason for the American semiconductor industry's troubles is that Japanese firms have continued to expand their facilities even during recessionary periods. In 1974-75, following the first "oil crisis," the industrialized world was hit by a severe recession. At the time, American semiconductor makers cut their plant and equipment investment in half. Japanese chip manufacturers, on the other hand, were not reluctant to invest. As a result, when the economy revived and customer demand began to pick up, Japanese firms were able to capture a larger share of the market.

A *third* reason is that Japanese tariffs on imported semiconductors are higher than those of the United States. The American tariff on semiconductor products is only 5.6 percent, while Japan imposes a duty of 10.1 percent. Compared with the 17 percent tariff of the European Economic Community, Japan's rate may not seem really high, but in response to demands from the United States, Japan lowered her tariff on semiconductors to 4.2 percent, beginning in 1982. At the same time, America also lowered her own tariff to 4.2 percent.

The *fourth* reason cited by American semiconductor firms for their predicament is that the Japanese government subsidizes its domestic semiconductor industry. The American chip manufacturers complain that the U.S. government only subsidizes old industries like steel and automobiles, while giving no support at all to new industries like semiconductors. The American firms appealed to the government in May, 1981, for federal tax credits for research and development expenditures.

Only a month later, in June, NEC Corporation (Nippon Electric) of Japan announced that it would build a huge semiconductor manufacturing plant in California. The factory at Roseville, just north of Sacramento, represents a $100 million investment by NEC, and opened in early 1983. The new plant will reach a peak monthly production level of 75,000 to 80,000 semiconductors by 1986, and NEC hopes to gain a 10 percent share of the market in the United States by then. Two of the semiconductors to be produced by the Roseville factory, one of the largest in the country, will be a 64K RAM chip and a 128K ROM (Read-Only Memory) chip.

American industry analysts had expected that Japanese semiconductor makers would someday begin opening production facilities in the United States, but the scale of NEC's planned factory apparently surprised even the experts. Thus, in the field of computers — currently the most substantial and most important of the "third wave" technologies listed by Toffler — America is in the process of being defeated by Japan. Whether the recent crash projects by the U.S. government can match the Japanese effort or not is yet to be decided. The fifth and sixth generation battle for artificial intelligence computer superiority has just begun.

Why Can't America Compete?

Since the end of the second World War, the United States had led the world in practically every field. Thus, being overtaken by Japan while no one was looking, as it were, has been a major shock to the American people. Any number of popular magazines have taken up this issue in recent years. The weekly magazine *Time* featured a cover story entitled "How Japan Does It," which described Japan as "the world's toughest competitor." The picture on the cover of that issue is a masterpiece. It is drawn exactly in the style of a Japanese woodblock print of the feudal era, depicting a traditional Kabuki actor grasping a paper umbrella. This particular gentleman of Japan, however, also carries a 35mm camera around his shoulder, wears a digital watch on his wrist, clutches a pocket calculator, a set of Japanese car keys, and an attache case in one hand, and holds a golf club in the other hand along with his umbrella.

The following summary of *Time*'s cover story appears in the table of contents for that issue (March 30, 1981): "First it was ships and steel. Now the Japanese have conquered markets for autos, televisions, stereos, and watches. How have they managed to become the world's toughest competitors?" The cover article was accompanied by a number of color photographs, including an employee's wedding ceremony at Matsushita Electric's company club, and row upon row of Datsun automobiles unloaded at an American seaport. The article, which deals primarily with Japanese management, makes use of a number of actual Japanese words, as in the following passage:

> For all their cross-cultural borrowing, the Japanese have remained aston-ishingly unchanged. One of the most important of their native charac-teristics is a willingness to achieve consensus by compromising. Asian Scholar Edwin Lee of Hamilton College suggests that a clue to this might be found in the Japanese word *ie*, a concept that can be inter-changeably applied to everything from self to home to family. A person is an extension of his immediate family members, his company, his community, and his nation as a whole. All are bound together in an encompassing common purpose.

How did a concept like *ie* originate? *Time* goes on to explain:

> Japan feels itself to be a "family" because in a real sense nearly every-one has at least some voice in running society. No matter what the whole group — from the smallest upstart enterprise to the largest multi-billion dollar multinational — nothing gets done until the people in-volved agree. The Japanese call this *nemawashi* (root binding). Just as a gardener carefully wraps all the roots of a tree before he attempts to transplant it, Japanese leaders bring all members of society together be-fore an important decision is made.
>   The result is an often tedious, and sometimes interminable, process of compromise in the pursuit of consensus. But in the end the group as a whole benefits because all members are aligned behind the same goal.

Less than one month after *Time*'s cover story appeared, a book was published with the title *Theory Z: How American Business Can Meet the Japanese Challenge*. Its author is William G. Ouchi, a 37-year-old professor at the school of management of the University of California at Los Angeles, who has been studying, since 1973, the management techniques used by Japanese companies. This book became a sudden best-seller in America. The *New York Times* commented:

> Books on management usually appeal only to a specialized audience. Typically, they sell a few thousand copies at most to scholars or business executives. The author's obscurity is undisturbed and his remuneration is paltry. Thus, according to the normal dictates of publishing, Mr. Ouchi's book should have been ignored. But not today.

The initial press run of 50,000 copies sold out immediately, and the Japanese firm CBS-Sony purchased the Japanese translation rights for a record-breaking sum of $110,000. "Theory Z" is the name given by Ouchi to Japan's style of management techniques. As he explains, this is an allusion to the popular classification of management techniques as "Theory X" and "Theory Y," devised by Douglas McGregor, a professor at the Massachusetts Institute of Technology (M.I.T.). According to McGregor, some managers hold the view that "people are basically lazy and irresponsible, so I have to watch them constantly." Other managers, however, operate on an opposite theory that "people are basically diligent and responsible, so all I need to do is assist and encourage them." McGregor refers to the management technique based on the "lazy" view of human nature as "Theory X," while the technique based on the "diligent" view is dubbed "Theory Y." Ouchi's "Theory Z" goes beyond these two techniques. In companies that follow "Theory Z," lifetime employment is held up as the ideal. For the first ten years after a new employee enters the company, his performance is not evaluated. As a consequence, he feels no need to compete with his fellow employees, so work gets done with everyone cooperating. Moreover, a company of this kind tends not to foster specialists, but rather generalists or "allrounders," who become familiar with all aspects of the company.

Ouchi's book introduces a series of fictitious employees named Sugao, Fujioka, and the like, who work at the Mitsubeni Bank and other imaginary firms. He even describes how Japanese white-collar workers, instead of returning directly home from their offices, often stop to browse in a bookstore, perhaps, or to play the pinball game called *pachinko*. In June, 1981, another book on this subject appeared, with the title *The Art of Japanese Management: Applications for American Executives*. Its two authors are Richard T. Pascal, a lecturer at the Stanford University graduate school, and Anthony G. Athos, professor of business administration at the Harvard Business School. In this book, such features of Japanese company life as the "*sempai-kohai* (senior-junior) relationship" are cited as constructive forces. *The New Republic*, in its issue for the week of June 27, 1981, used a review of these two books to introduce the term "samurai management." The cover of the magazine featured a striking picture of an American executive brandishing a samurai sword.

The situation was the same in 1982, 1983, and is again being repeated in 1984. The fad over Japanese techniques continues. Other magazines and newspapers carried frequent stories describing the "Japanese miracle," and discussing such Japanese institutions as lifetime employment, promotion by seniority, quality control, and bureaucratic guidance. These were seen as keys to Japan's success. What these articles pointed out may have been partly on target. But as a Japanese, I feel rather uncomfortable, rather embarrassed, when Americans praise the Japanese business world for the distinctive features mentioned above. *Japan as Number One, The Japanese Mind*, etc., all leave me slightly confused.

For example, the *ie* system described by *Time* was abolished by the American occupation authorities after the second World War, at least in a legal sense as it applies to the family. The reason given by the occupation was that Japan's patriarchal family system was a holdover from the feudal period, and ran counter to the ideals of democracy.

Lifetime employment and seniority-based promotion have until very recently been severely criticized by the Japanese themselves. A popular view among younger Japanese today is that it is foolish to devote one's entire life to a single company, and that like Americans, Japanese workers ought to keep changing jobs and find new fields of activity that satisfy them personally. Books with titles like *Why You Should Change Your Job* keep appearing. A view often expressed is that working for other people is pointless, and that it is better to give up a salaried job in favor of independence by starting one's own business or opening a shop.

The concept of the "*sempai-kohai* relationship" is constantly coming under attack for ignoring the abilities, efforts, and achievements of the individual employee. Under the "*sempai-kohai*" system, an employee's relative ranking within the company is determined by the number of years he has worked there, and his age. Japanese often remark that in America, employees of outstanding ability, no matter how young, are constantly being singled out to become managers and directors, and that only when younger people have leadership positions in society, can that society develop in a dynamic manner.

Thus, institutions that at least until very recently were seen by some Japanese as outmoded and harmful are now being praised by Americans as the keys to success. Many books and articles on Japanese management techniques will, no doubt, continue to appear in America. But in my own view, most writings of that kind tend to overlook one extremely significant factor. The factor is what I call the "anti-science disease." It might also be termed the "anti-technology disease."

The technology of the past was slowly developed and refined by artisans over long years of personal experience. Nearly all of today's technology, however, is based upon scientific knowledge. In my view, science and technology are inseparably bound together. As a consequence, the anti-science disease and the anti-technology disease spring from the same basic causes, and their symptoms are largely identical. For that reason, I will consider the anti-technology disease to be included in my discussion of the disease of anti-science.

This disease first appeared in America in the latter half of the 1960's, and had reached its most virulent stage by the early 1970's. Anti-science proved to be contagious, and spread like an epidemic throughout the world. Japan, too, came down with the sickness and was gravely ill for a time. But Japan caught the disease some years later than America, and was several years ahead of America in recognizing the dangers of this sickness and taking steps to deal with it. America was late in trying to halt and cure the epidemic. The severity of America's attack of anti-science, compared with Japan's less serious case, is in my opinion the main reason that Japan has overtaken American industry in so many areas.

Americans today are gazing across the Pacific Ocean to Japan, in the belief that they can somehow ferret out Japan's "secret of success." This feeling is completely understandable. But to my Japanese eyes, Japan does not seem to possess any particular "secret." Whatever the advantages of customs like lifetime employment and seniority-based promotion, such practices are unlikely to take root in American society. Before Americans try to discover Japan's "secret," perhaps they need to diagnose accurately the advanced case of anti-science disease from which they themselves are suffering, and find an appropriate cure.

The problem is not "How Japan Does It" — rather, the question should be "Why Can't America Do It?" My interpretation of the problems is that America has come down with a severe case of anti-science and is too ill to push forward with adequate research and development in a number of vital areas. While America has been lying on her sickbed, Japan has moved ahead of her.

II
AMERICA'S SCIENCE ALLERGY

I use the term "anti-science disease" as a general term for any antipathy, hatred, or hostility toward science and technology. Such attitudes can also be expressed in the form of violent acts of resistance. At times, this has become a political issue. President Nixon, in an attempt to seek out views on science and technology policies for the 1970's and beyond, announced on October 6, 1969, the formation of the Presidential Task Force on Science Policy. The chairman of this group was Dr. Ruben F. Mettler, executive vice-president of TRW, Inc., while the distinguished list of members included a university president, the chairman of a corporate executive committee, the president of the National Academy of Sciences, and the director of one of the National Laboratories. In April, 1970, the Task Force submitted to the President a report entitled *Science and Technology: Tools for Progress*. One section of that report was called "Anti-Science Attitudes." In that portion of the report appeared this passage:

> The rapid rise of attitudes disdainful of science and technology, and the disillusionment of many young people with science and technology is of grave concern. The sources of these attitudes include deficiencies in the application of science and technology which should in fact be criticized and should be corrected. . . .The sources of the shift in attitudes toward science and technology also include widespread lack of perspective and understanding of their nature and role in past and future improvement in the human condition.

The Task Force went on to propose the following approach to dealing with "anti-science attitudes":

> The public and its elected representatives must have a better grasp of both the limitations and the promise of science and technology. Priority should be given to presenting this complex matter to the public in a balanced and understandable fashion. The responsibility for achieving this understanding starts with the Executive and Legislative branches of the Federal government and spreads to include state and local government, universities, business and professional organizations, and other private institutions in positions of leadership.

The recommendations of that Presidential Task Force, however, appear to have had little effect. I say this because America's anti-science disease subsequently showed no signs of abating. On August 16, 1971, Dr. Franklin P. Huddle, of the Science Policy Research Division of the Congressional Research Service at the Library of Congress, submitted to the U.S. Senate Interior Committee a report entitled *The Evolution and Dynamics of National Goals in the United States*. Huddle, too, felt compelled to include in his report a section called "The Emergence of Anti-Science Philosophy," in which he discussed the hostility toward science and technology currently common among intellectuals. He wrote: "The imperfections of the American technical culture and even the culture itself are beginning to come under attack. It is worthwhile to examine the basis for this new trend, and the anatomy of its manifestations." At one time, Huddle said, technology played a major role as a "tool" of the nation. He recalled that:

> The role of technology in World War II was decisive. In implementing the goal of containing communism, technology occupied a central place: development of super-weaponry, devising counterinsurgency systems, and extending technical assistance in the foreign aid program. Technology became the goal itself after 1957 in the space program.

Later on, however, technology was toppled from the high position it occupied in America's national goals. As Huddle described it:

> Then, when the nation became concerned over quality of the environment, it discovered that the impairments of quality were attributable in great part to the very technological innovations that had been so eagerly embraced in past years. It is no accident that much recent legislation to improve the environment deals explicitly or indirectly with the regulation of technology, and less with the exploitation of new technology.

Huddle went on to point out:

> It is noteworthy that the discussion of national goals by a Commission formed by President Eisenhower [to be discussed later on] regarded technology as a hopeful means of enlarging the capacity of the individual and achieving national goals at home and abroad, whereas President Nixon's Goals Staff was more concerned with the ills resulting from past technology.

Huddle then made the prophetic statement that "the challenge to technology — and to the basic sciences that expand its reach — may go deeper." That is exactly what happened.

Doubting Science

The anti-science disease is certainly nothing new. Looking back into history, we can discover even in ancient Greece a tendency to belittle or attack science and technology. Technology was particularly despised. The history of science began in the region of Ionia, on the eastern shore of the Aegean Sea, around 2,500 years ago. The philosopher Thales of Miletus, who lived about 580 B.C., is said to have been one of the first to teach that events could be explained without recourse to divine causation. This view can be considered the beginning of scientific thinking. Before the time of Thales, men believed that everything in the world depended upon the will or whim of the gods. A flood, a drought, the death of a child — everything happened because a deity wished it so.

But Thales of Miletus and his followers believed that natural events followed rules, and that it was possible to discover those rules. They banished the gods, as it were, and preached the importance of observation and experimentation. This way of thinking grew more and more influential as it spread to various branches of learning. The writings left by the physician Hippocrates of the Ionian island of Cos (fifth and fourth centuries B.C.) clearly express the thinking of the scientists of that period. It was then widely believed that the disease of epilepsy was of divine origin — that it occurred when a god took possession of a human being. But Hippocrates wrote:

> It seems to me that the disease is no more divine than any other. It has a natural cause, just as other diseases have. Men think it divine merely because they do not understand it. But if they called everything divine which they do not understand, why, there would be no end of divine things.

Hippocrates banished the gods from the study of medicine, and tried to find the natural causes of illness. Technology also made great progress, along with science, during that same period. Enormous stone temples, massive walls, and long aqueducts were constructed. On the Ionian island of Samos in the Aegean Sea, engineers as early as the sixth century B.C. bored a tunnel 3,300 feet long through a mountain in order to carry an aqueduct. The work, which took 15 years to complete, was carried on from both sides simultaneously, and the two holes met almost perfectly inside the mountain. That feat is an indication of the strides that had by then been made in the technology of surveying and civil engineering. Such techniques put to practical use the discoveries of pure science like geometry. A genius of technology named Theodorus, from the island of Samos, is said to have invented a lock and key, a level and a square used by carpenters, a lathe, and a method for casting bronze.

Ionian science and technology gradually spread to the nearby shores of Greece, then to Italy, Egypt, and elsewhere in the ancient world. But only 200 years or so after the time of Thales of Miletus, people were already belittling and attacking technology. The historian Xenophon (431-350 B.C.) displayed that attitude when he said:

> What are called the mechanical arts carry a social stigma and are rightly dishonored in our cities. For these arts damage the bodies of those who work at them or who act as overseers, by compelling them to a sedentary life and to an indoor life, and in some cases, to spend the whole day by the fire. This physical degeneration results also in deterioration of the soul. Furthermore, the workers at these trades simply have not got the time to perform the offices of friendship or citizenship. Consequently they are looked upon as bad friends and bad patriots, and in

some cities, especially the warlike ones, it is not legal for a citizen to
ply a mechanical trade.

About 150 years after Xenophon wrote those words, the great technologist
Archimedes (287-212 B.C.) discovered the principle of the lever. The use of a long
stick as a lever to move a heavy object had been known from antiquity, but it was
Archimedes who first described the fixed mathematical relationship between the
length of a lever and the amount of force it can exert. "Give me a place to stand," he is
supposed to have said, "and I can move the world." He discovered a method for
determining whether an object was made of pure gold, without harming the object,
and he invented a helical pump that is still called the Archimedean screw.

Archimedes himself, however, left hardly any writings describing his technological inventions of this kind. In his era, it was felt that persons of high social
status should not busy themselves with devices to make or move things. Such
"mechanical arts" were fit only for slaves and laborers. Since Archimedes belonged
to a family of high rank in Syracuse, he found it an embarrassment to be interested in
the mechanical arts. He himself wrote about only one of his mechanical devices: a
contrivance that modeled the movements of the sun, moon, and planets. Since this
machine did not actually do any useful work, Archimedes found no reason to be
ashamed of it.

This attitude of ridicule and hostility toward technology eventually developed into
a similar feeling about science in general. The science and technology that had been
born in Ionia eventually declined beyond recognition. It was not until the sixteenth
century in Europe that life was again breathed into the scientific tradition. Through
the efforts of Copernicus and Galileo, the scientific way of thinking slowly revived.

The progress of science and technology since that time, however, has not been
without obstacles and interruptions. For example, there is the case of Antoine
Lavoisier, who is known as "the father of modern chemistry." It was he who
discovered that combustion is supported by the oxygen in the earth's atmosphere, and
he also formulated the law of conservation of mass. He compiled the first listing of
chemical elements, and in 1789, summing up his entire body of work, he published
his *Traite elementaire de chimie*, considered the first textbook of modern chemistry.
But in that same year the French Revolution broke out. Lavoisier had accepted an
official post under the royal government, collecting taxes on such goods as salt and
tobacco. An official like Lavoisier was a kind of contractor. He had to collect a certain
amount of taxes for the government, but if he was able to collect more than his quota,
any additional money became his. Lavoisier not only collected taxes, he even married
the daughter of a high official in the tax administration.

With the money he earned as a tax agent, Lavoisier built a superb laboratory, and
began laying the foundations of modern chemistry. However, the extremists who had
taken part in the revolution began to make reprisals against officials like tax
contractors. Lavoisier, too, was quickly arrested. He pleaded his case in these terms:
"I am a student of chemistry. I am not involved in politics in any way. All the money I
have earned as a tax collector, I have spent on my research. I have done nothing
unlawful. I am a scientist." But the captain of the soldiers who had arrested him
declared brusquely: "The Republic has no need for scientists." On May 2, 1794,
Antoine Lavoisier was beheaded by the guillotine. An astronomer of the same era,
Count Lagrange, uttered this famous comment about Lavoisier's execution: "A

moment was all that was necessary to strike off his head, and probably a hundred years will not be sufficient to produce another like it."

Around that time, in England, machinery-smashing movements sprang up. As scientific thought had revived, technology had also begun to push forward once again, and a cotton-spinning machine had been invented in England in 1764. The next year, James Watt succeeded in greatly improving the steam engine. As a result of such inventions, the Industrial Revolution was underway in Britain. But not everyone welcomed this kind of progress. In January, 1768, new mechanical looms were attacked by workers who used hand-operated looms. With the new machinery, one worker was able to do the work of six using hand looms, and the weavers believed that they would lose their jobs. Also in 1768, a sawmill in London was attacked by 500 angry workers because it used mechanical saws. These workers, who cut lumber by using hand saws, were afraid that they would be robbed of their livelihood.

The Industrial Revolution began in England around 1770, but it was not until later, between 1811 and 1816, that the violent Luddite movements developed in that country. These were a series of campaigns aimed at physical destruction of the new spinning machines, power looms, and stocking frames. On March 11, 1811, a protest meeting in the marketplace at Nottingham led to the smashing of six machines. That was the spark that ignited a conflagration. Between March 16 and 23, rallies were held in many villages and towns, and more than a hundred machines were destroyed. The violence spread throughout Britain.

The name "Luddite" was given to the anti-machinery movement because, in December of 1811, a threatening letter signed "Ned Ludd" was sent to a factory owner. The name is apparently a fictitious one, however — no actual Ned Ludd was ever identified. Subsequently, many similar letters were sent, often signed "General Ludd" or "King Ludd." Often the Luddites blackened their faces, carried pistols, and used piles of hay and straw to set fire to buildings. Some factory owners who resisted the Luddites were murdered by them. A considerable number of bandits and highwaymen apparently took to calling themselves Luddites. With those disreputable characters thrown in for good measure, the movement persisted until about 1816. When Hargreaves and Arkwright of Nottingham had invented their spinning machinery, they had received popular praise, but now the hostility of the Luddites forced them to leave Nottingham.

Another major anti-science movement appeared about 100 years later, in the 1920's in Germany, one of the countries defeated in the first World War. In that war, for the first time, poison gas had been employed, and aircraft were first used as instruments of death. Science and technology must have inspired a vague sense of fear in the public. Many intellectuals complained that science and machines were leading civilization down a dangerous road. The scientific way of thinking was attacked head-on by some. They asserted that scientists broke down objects and phenomena into their constituent parts and elements in order to study them in an analytical manner, but that this so-called "elementalism" ignored the totality of the object. The scientific, rationalist mode of thought, these critics charged, led only to egoism and materialism. Irrational intuition, some philosophers insisted, should be of primary importance.

Mathematics, which uses logical proofs, also came under attack. Physicists, too, who rigorously pursued the relationship between cause and effect, became targets for hostility. We must abandon technology, said some, advocating a return to handicrafts. This anti-scientific trend in Germany eventually spread across Europe and the Channel to Britain. There, in 1927, the Anglican Bishop of Ripon called for a moratorium on scientific research. "For the next ten years," he said, "let us halt all scientific research completely."

This proposal attracted notice in America. Newspapers and magazines featured the Bishop's plea, and it became a topic for conversation everywhere: "The Great War was a tragedy. If science is allowed to go on like this, the next war will be a catastrophe." "Machines rob mankind of its humanity. The automobile industry may be able to provide jobs for tens of thousands of people, but assembly-line workers can't feel the joy of labor. Today's blue-collar worker has none of the pride in his work that the old-time craftsman had." "Education in the schools has been poisoned by 'scientism.' We teachers are simply prostituting ourselves to scientific knowledge, and we aren't able to provide an education for 'the whole man' anymore." "Modern man may know far more things than Socrates did — but we aren't any wiser than Socrates, or any nobler in character." "America ought to give up technology and go back to the good old days."

A great many intellectuals loudly delivered themselves of remarks like these. The scientific way of thinking gradually weakened, and in its place, pseudo-sciences like astrology began to grow very popular. A magazine in Germany published a special issue on astrology, which included a cartoon about astrology and astronomy. The cartoon showed an astronomer, hard pressed for research funds, who becomes an astrologer, thereby earning a huge fortune. He then rebuilds his observatory facilities, and goes back to his astronomical research. In that period, too, there were people who opposed advances in technology on the grounds that the new technology and automatic machines caused unemployment. This belief was reinforced by the great worldwide depression that began in 1929. The wave of anti-science and anti-technology sentiment left its mark right down to the start of the second World War.

Technology Feared

This anti-science disease has been with us since antiquity. Yet this attitude, now popular throughout the countries of the free world, is prevalent to a degree unprecedented in history. On the other hand, the causes of the current opposition to science and technology are more numerous, and more complex, than at any previous time in history. Today's anti-science disease, like its counterpart in ancient Greece, seems to have begun with a hostility toward technology.

In 1948, people had faith in the usefulness of science and technology, but in the same year the Swiss architectural historian Siegfried Giedion published a book titled, *Mechanization Takes Command*, which emphasized the harm caused by mechanization. This book of more than 700 pages traced the history of how mechanization was applied to everything from farms and bakeries to slaughterhouses and toilets. In his book Giedion wrote:

> What does mechanization mean to man? Mechanization is an agent, like water, fire, light. It is blind and without direction of its own. It must be canalized. Like the power of nature, mechanization depends on man's capacity to make use of it and to protect himself against its inherent

perils. Because mechanization sprang entirely from the mind of man, it is the more dangerous to him. Being less easily controlled than natural forces, mechanization reacts on the senses and on the mind of its creator.

According to Giedion, mechanization fragments human labor. With the assembly-line system, originated in America, the individual worker never completes a single product by himself. A worker may be assigned to duties on a particular part whose ultimate purpose he does not even know, or he may be required to spend an entire day, for example, tightening one particular screw each time a part passes the work station. Or, to take the case of the consumer, mastering the operation of machinery manufactured products is becoming more and more difficult. When an automobile's engine suddenly stops running, the car's owner probably has no idea which part is defective. He peeks under the car hood, but looks away in confusion.

"Man is being defeated by his own tools," said Giedion. "Future generations will perhaps designate this period as one of mechanical barbarism, the most repulsive barbarism of all." Giedion was born in 1888 and lived during the "era of anti-science" attitudes that lasted from about 1920 to 1940. He was undoubtedly sympathetic to this mood and when the second World War ended he decided to revive the anti-science philosophy he knew from the prewar period.

In 1954, France's Jacques Ellul wrote a book called *The Technological Society*, which described "the tragedy of a civilization dominated by technique." According to Ellul, the first characteristic of modern "technique," as he called technology, was rationality. For the sake of rationality, everything must be systematized, occupations must be fragmented, standards must be created, and norms had to be established. A second characteristic, he says, is artificiality. Technology runs counter to nature, and produces an artificial world, one that is fundamentally alien to nature. Since technology breaks down the natural world, eradicates it, and forces it into submission, it may well become impossible for the natural world to be restored, to co-exist with technique.

"We are rapidly approaching the time," asserted Ellul, "when there will no longer be any natural environment at all." In his view, man reached a point where he could no longer control his technology. He has lost control over the milk he drinks and the bread he eats. He is unable to run the giant factories or the systems of transport and communication. With technology, harmful "side effects" were inevitable, and additional new technology would be needed to overcome those effects. For example, Ellul writes:

> To make housework easier, garbage-disposal units have been put into use which allow the garbage to run off through the kitchen sinks. The result is enormous pollution of the rivers. It is then necessary to find some new means of purifying the rivers so that water can be used for drinking. A great quantity of oxygen is required for bacteria to destroy these organic materials. And how shall we oxygenate rivers? This is an example of the way in which technique engenders itself.

Ellul insists that "technique cannot be otherwise than totalitarian." Technology is gathered into the hands of the state, and the state becomes totalitarian. This work by

Ellul, initially published in French, attracted little attention at the time. It appeared in 1954, but at that time neither America nor Europe had yet been afflicted with the anti-science disease. In 1961, the head of an American publishing firm noticed Ellul's work, and had it translated into English and published (1964). That was just about the time that America was coming down with its anti-science illness. As a result, the English-language edition became an instant success, and contributed greatly to the severity of the anti-science disease in America.

Man is influenced by "ideas." Man constantly acts in accordance with the concepts and notions that lodge themselves in his mind. A remarkable example of this is the experience of a Japanese, Shoichi Yokoi. As a sergeant in the Japanese army during the second World War, he was stationed on the island of Guam. There were about 20,000 Japanese soldiers on that island, but on July 21, 1944, when 54,000 U.S. Marines came ashore, the Japanese were defeated. Perhaps 10,000 of the Imperial Army soldiers fled into the jungle in the northern part of the island. Shoichi Yokoi was one of them. In spite of repeated attempts to persuade him to surrender, Yokoi continued to hide in the jungle, living there for 26 years and five months after the end of the war. It was not until January 24, 1972, that he was finally discovered and captured by two islanders.

One very simple "idea" had kept Yokoi from coming out of the jungle during those years: Japanese soldiers had all been strictly taught that they must never suffer the shame of being taken a prisoner alive. After his capture, Yokoi faced a group of Japanese reporters. Had he known that the war was over? He replied, "I found out the year after the war ended, when I picked up a Japanese newspaper printed in Guam. Also, I heard announcements over loudspeakers ordering Japanese stragglers to surrender." Why had he not come out of the jungle and given himself up? "In the army, we were taught to 'fall like the cherry blossom, with the spirit of the samurai.' So I was supposed to die the way the cherry blossom withers."* Did Yokoi want to return to Japan? He said, "I am ashamed to return home after being captured. But I would like to go home just once."

The idea that "machines are evil" and that "technology leads to totalitarianism" exerted a powerful influence over the countries of the free world during the past twenty years. Like the ancient Greek notion that "the mechanical arts carry a social stigma and are rightly dishonored," this attitude prodded society in a single direction.

I do not intend to discuss the views of Giedion or Ellul in greater detail, or to argue whether those opinions are correct or not. What they say certainly contains much that deserves to be listened to. What we need to do is pay close attention to the side effects that are inherent in our technology, such as the social dislocation caused by the elimination of jobs. Still, how objective is Ellul's view that technology becomes totalitarian as it advances? Compared with today's Japan, prewar Japan had a far lower level of technology, and ordinary Japanese people were without electric washers, refrigerators, and television. Yet in those days, the military controlled a totalitarian government. Citizens had no freedom of speech or assembly, and young people were drafted into the army and sent off to war. Compared with that era, Japan today is far more developed technologically. Japan is also immeasurably more free, more democratic, than in the prewar period. Thus, technological progress does not necessarily lead inevitably to totalitarianism. Moreover, just as in the days of the

---

* In Japan, the cherry blossom, whose petals fall at the peak of their bloom without withering or fading, symbolizes the "Japanese spirit" of the samurai, who was supposed to die gloriously in battle, with his honor unsullied.

Luddite movement in England, there are still those who oppose progress in technology on the grounds that new techniques cause a loss of jobs. One example of this can be found in the newspaper industry. Since Johannes Gutenberg invented printing with movable type in the 1450's, newspaper publishers, like other printers, have continued to use movable type cast from lead alloy. But the time has now come to say good-bye to the Gutenberg method. With the development of the computer, newspapers can now utilize letter images generated by a computer and reproduced on film, in a so-called "cold" printing process using offset lithography. Newspaper publishers in the advanced nations began to introduce this new technology around 1974. It followed logically that the traditional printing craftsmen who worked with movable metal type were no longer needed. According to an authority like Bruce Gilchrist, director of the Center for Computing Activities at Columbia University, over a four-year period (1974 to 1978) three major daily newspapers in New York City reduced their number of typesetters by 31 percent, 39 percent, and 43 percent, respectively. Gilchrist and his associates interviewed 44 typesetters who had lost their jobs, and traced their subsequent lives. One 58-year-old man who had worked for 30 years as a printer received severance pay of only $4,000, since he had been employed by the newspaper company for only one year before being terminated. He had a wife and one son. He was still paying off the mortgage on his house. Although he had a certain amount of money in addition to the $4,000, he was despondent. His wife told the interviewer, "I can't just leave him here at home alone and go out to work. I feel as if he would just fall apart. . . ." A few months later, his wife was admitted to a mental hospital. The next year, he tried to find a job at a printing company, but the pay offered by such firms was about $150. That was about one-third of what he had been earning earlier. His self-respect was wounded and he took a better paying job as a telephone switchboard operator. His income, however, was still only half of what it had been at the newspaper.

Others studied by Gilchrist had their share of hardship, too. They had all been labeled "newspaper typesetters," hence ordinary printing companies were very reluctant to hire them. Newspaper publishers with similar plans to introduce new printing technology also had no desire to hire any more typesetters. For someone who is healthy and wants to work, unemployment is truly a painful experience. In August, 1978, the typesetters on New York City's three big dailies went on strike to protest the introduction of computerized typesetting. As a result, the *New York Times* and the two other papers were unable to publish for 87 days. A few months later, on November 30, 1978, the *Times* of London announced in an editorial notice that "there will be an interval," and thereupon ceased publishing. The reason? The paper's labor union had gone on strike against computerization. The *Times* did not reappear on the streets until a year later. Machines and technology are altering the nature of people's work. Even Giedion, the anti-mechanization advocate, concedes that "no doubt, mechanization can help eliminate slave labor and achieve better standards of living. Nevertheless. . . ."

According to an analysis by Marc Porat, the types of work being performed by the population keep changing with time. Figure 7 is a chart prepared by Porat showing these changes in the American labor force. Until the start of the twentieth century, agricultural labor represented nearly 50 percent of the total work force but today it is only 4 percent of the labor force. During the decade between 1930 and 1940, the ratio of industrial workers to the total work force reached about 40 percent. After 1950, the

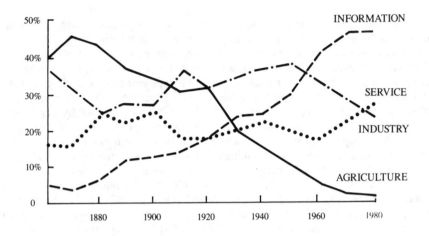

Figure 7   THE FOUR SECTORS OF U.S. LABOR FORCE

number of industrial workers began to drop, while the percentage of workers in service industries and the information industry increased. This change in the composition of the work force closely corresponds to the change from the "first wave" to the "third wave," as described by the writer Alvin Toffler. Toffler says the "third wave" — the next era — will continue to expand steadily, with the increasing use of microcomputers. The American scientific journalist Colin Norman, in an article titled "The New Industrial Revolution," analyzes in detail how microelectronics will deprive people of work: the production of digital watches using microelectronics in America and Japan caused a loss of 46,000 jobs in Switzerland in the 1970's. During this period, seventeen Swiss watchmaking firms went bankrupt. The reason? A tiny piece of semiconductor circuitry replaced several hundred moving mechanical parts.

According to the 1975 annual report of National Cash Register, the amount of labor needed to manufacture one electronic cash register is only 25 percent of that needed for one of its mechanical or electromechanical counterparts. The seven Japanese companies that make color television sets increased their total annual production by 25 percent between 1972 and 1976. Yet the total number of production-line workers they employed dropped from 48,000 to 25,000 during that same period. In light of figures like these, people who are hostile toward machines and technology, charging that progress in technology causes unemployment, have a substantial basis for unhappiness. Nevertheless, the question of whether or not unemployment will result from technology depends upon the social system. It is society's responsibility, not that of technology as such. My newspaper, the *Asahi Shimbun*, did away with movable metal type on September 23, 1980, when we converted to a system of computerized composition. We were a little behind American and British newspapers in making this change, in part because of the complexity of the Japanese writing system. The English alphabet has only 26 letters, while a Japanese-language newspaper uses around 3,000 different written characters. Thus, my newspaper took longer to devise and implement its system, but with the introduction of computer technology, about 300 typesetters were no longer needed. Not one of the *Asahi Shimbun's* typesetters was fired, however. All were trained in new jobs and relocated in other areas of the newspaper. The *Asahi Shimbun* was not hit by any strikes like those at the *Times* of London and the *New York Times*.

As I stated in Chapter I, Japan is the world's most advanced country in the use of industrial robots. But hardly any Japanese workers have been dismissed because of robots. The International Labor Organization, in a report published on June 1, 1981, analyzed the Japanese attitude toward robots. As an example, the report chose the Nissan Motor Company's plant at Zama, Japan, which produces Datsun automobiles. With the installation of 75 assembly-line robots in that factory, not one production worker is needed. The unneeded workers, however, have all been placed in other jobs at Nissan and its affiliated companies, and no one has been fired or laid off. "The introduction of robots," says the report, "has proceeded under an agreement with the labor unions." Many Japanese companies, of course, adhere to the lifetime employment system, so that even when an employee's job is eliminated, the firm cannot simply fire him. This fact can be seen as tending to appease and moderate any hostility toward technological advances. In America and the European countries, however, fear of unemployment is a major factor in "anti-technology" thinking. In those countries, when company executives try to introduce new technology into the workplace, the workers resist strenuously. This is only natural, since they fear losing their jobs.

Colin Norman, however, notes that:

> Every expert who has studied the potential employment impact of microelectronics has reached the same conclusion: more jobs will be lost in those countries that do not pursue the technology vigorously than in those that do. The reason is that microelectronics will enhance productivity to such an extent that the industries that move swiftly to adopt the technology will have a competitive advantage in international markets.

As long as the principle of competition in the international market is operative, a country that falls behind in technological development will, like today's American automakers, end up with large numbers of unemployed workers.

The Science Counterculture

Hostility toward technology can easily develop into hostility toward science. In this connection, let us look at the writings of the American historian Theodore Roszak, one of the leading champions of anti-science thought. He was born in Chicago in 1933, studied at the University of California, and received a doctorate from Princeton University in 1958.

In 1969, Roszak published a book entitled *The Making of a Counter Culture*, an instant best-seller, selling 400,000 copies by 1972. In this work, he attacks today's "Western civilization," by concentrating his fire upon science. Roszak insists that modern civilization must be resisted, and that a new civilization must be brought into being. With the success of this book, the use of the term "counter culture" became commonplace in the mass media. Writing in the journal *Daedalus*, Roszak referred to Victor Frankenstein, the hero of Mary Shelley's 1818 novel. The young medical student Frankenstein gathers parts of human bodies from graveyards and dissecting rooms, and builds from them a monster who is quite human in appearance. But the monster, estranged from human society, becomes violent, and in the end turns on his creator Frankenstein and destroys him. Roszak writes:

> Asked to nominate a worthy successor to Victor Frankenstein's macabre brainchild, what should we choose from our contemporary inventory of terrors? The bomb? The cyborg? The genetically synthesized android? The behavioral brain washer? The despot computer? Modern science provides us with a surfeit of monsters, does it not?

Thus Roszak places both technology and science on the same chopping block, as it were, and proceeds to hack away at both. Science obeys three great principles: objectivity, elementalism, and quantitativism. Roszak attacks all three. The first of these, the principle of *objectivity*, means that when investigating objects or phenomena and considering causes and effects, the scientist must eliminate as much as possible his own emotions or preconceived ideas. Ever since the days of Galileo, who awakened science from its long centuries of slumber, it has been asserted again and again that an attitude of objectivity is essential to discovering the laws and truths of the natural world.

The English philosopher and statesman Francis Bacon, who was a contemporary of Galileo, emphasized in his book *Novum Organum* the importance of objectivity. He pointed out that human beings are easily swayed by preconceptions and prejudices, and tend to interpret things in accordance with their own personal likings. As an illustration, Bacon tells the story of a certain temple in ancient Greece. The temple was hung with pictures of appreciative sailors who had been protected from shipwreck by praying to the gods before going to sea. A priest of the temple had proudly shown these portraits to a skeptical visitor. "Now tell me," the priest demanded, "after seeing this, do you not believe in the power of our gods?" "But where," replied the unbeliever, "are the portraits of those who were drowned at sea even after offering prayers here?" Bacon commented:

> And such is the way of all superstition, whether in astrology, dreams, omens, divine judgments, or the like, wherein men, having a delight in such vanities, mark the events where they are fulfilled, but where they fail, though this happen much oftener, neglect and pass them by.

Insisting that one must take careful note of even those instances that do not seem to fit one's theories, Bacon emphasizes the importance of objectivity in science. In *The Making of a Counter Culture*, Roszak has a chapter entitled "The Myth of Objective Consciousness," in which he attacks the objectivity of science. To Roszak, objectivity is nothing more than a "myth," arbitrarily created by society. He says that scientists and technologists, by designating their own views as "objective facts," have forced those views upon the general public. He insists that what is called "objectivity" is simply a tool used by the technological bureaucracy, the "technocrats," to exercise control over the public. But objectivity, Roszak maintains, alienates man from the natural world, and causes alienation within society itself as well.

The second principle of science is called *elementalism*, referring to the manner in which science analyzes and dissects the targets of its inquiry, reducing them to constituent elements. But things are complex and the methodology employed by modern science, to break things down into component parts, is sometimes called "reductionism." This philosophy of analyzing things into their component parts was advocated by Francis Bacon, but according to Roszak, it is incapable of comprehending a human or natural entity as a single mysterious whole. If man and nature are simply dissected into a given number of component elements, says Roszak, then they become objects of no value, which can lead only to philosophical nihilism.

The third principle, *quantitativism*, implies a desire to count everything that can be measured. Quantitativism in science uses measurement and observation to produce numbers and quantities, and compares them with other numbers and quantities, in an attempt to understand the essences of things. Lavoisier, for example, was able to discover oxygen because he could measure precisely the weight of the metals and volumes of air used in his experiments. Roszak, however, adds quantitativism to his list of targets. By measuring everything in terms of quantity, he asserts, scientists lose sight of quality. This way of thinking has been applied very broadly. For instance, the tests taken by schoolchildren are scored numerically, and students are then ranked and classified according to those ratings. But while such scores may indicate a child's success in learning a given subject, they give no hint whatever of the qualities of character.

Some would insist that the things that cannot be quantified are of great importance to mankind, and that we lose sight of the things when we try to quantify everything. This is a valid point, but if this particular criticism is carried too far, it can engender an attitude that prevents even measurable things from being measured.

After elaborating upon these criticisms of science and technology, Roszak advocates an alternative cognitive system, one that he calls "gnosis," which involves a holistic awareness of objects and phenomena of the natural world, including even their essential moral and religious values. According to Roszak, historic man viewed his existence in this fashion prior to the dissemination of the scientific method:

> Our science, having cut itself adrift from gnosis, contents itself to move along the behavioral surface of the real — measuring, comparing, systematizing, but never penetrating to the visionary possibilities of experience. Its very standard of knowledge is a rejection of gnosis, any trace of whose presence is regarded as a subjective taint. Value, quality, soul, spirit, animist communion were all ruthlessly cut away from scientific thought like so much excess fat. What remained was the world-machine — sleek, dead and alien. What is a monster? The child of knowledge without gnosis, of power without spiritual intelligence.

Theodore Roszak also bemoans the fact that both science and technology have fallen into the hands of narrow specialists. As science and technology have grown more complex and difficult, it has become virtually impossible for the layman to comprehend it. The general public has little choice but to believe what it is told by experts. As a result, science and technology have become effective implements for controlling the populace. Roszak writes:

> Expertise — technical, scientific, managerial, military, educational, financial, medical — has become the prestigious mystagogy of the technocratic society. Its principal purpose in the hands of ruling elites is to mystify popular mind by creating illusions of omnipotence and omniscience — in much the same way that the pharaohs and priests of ancient Egypt used their monopoly of the calendar to command the awed docility of ignorant subjects.

Roszak also mounts an all-out attack on the technocrats, giving the title "Technocracy's Children" to Chapter I of his book *The Making of a Counter Culture*. The word "technocracy" was coined by an American inventor, W. H. Smyth, and was popularized by the American writer Howard Scott in 1932. The world was in the depths of the Great Depression at the time. Streets were full of unemployed workers and numerous companies had gone bankrupt. In such a circumstance, Scott said, endlessly increasing productivity would lead to permanent unemployment and debt, even to the collapse of the capitalist system — unless the system was drastically revamped. The economy, Scott argued, had to be guided by a central planning mechanism that would match consumption to output. To keep the system independent of political manipulation, he said, it would have to be run by technicians. He called this social system a "technocracy," and claimed that under such a system, Americans could have a standard of living ten times as high as in 1929, by working only 16 hours per week.

Scott promoted his technocracy as an ideal social system, but in the hands of Roszak, technocracy has turned into a system for controlling the people by means of specialized knowledge. Roszak says that technocracy is a social system that emerges when a society becomes highly industrialized, and that it is just such a system that characterizes America today. In America, the rulers seek the views of technical experts in order to justify their own actions. The technical experts justify their own opinions by saying they are based on scientific knowledge. In the end, they fall back upon the "authority of science." To quote Roszak:

> The general public has had to content itself with accepting the decision of experts that what the scientists say is true, that what the technicians design is beneficial. We arrive, at last, at a social order where everything from outer space to psychic health, from public opinion to sexual behavior is staked out as the province of expertise. The community dares not eat a peach or spank a baby without looking to a certified specialist for approval — lest it seem to trespass against reason.

The "counter culture" cult was supposed to overturn this evolving technocracy. Others went on to expand upon Roszak's views, arguing that science and technology were servants of "the System," and are therefore evil. In 1973, for example, a Japanese biologist named Atsuhiro Shibatani published a book, *The Theory of Anti-Science*, which introduced "anti-science" philosophy into Japan. In this work, Shibatani discusses the question of science's relationship to "the System." Shibatani wrote:

In its most extreme form (and unfortunately this is not rare), science has blatantly served some particular authority or source of wealth, plunging the general public, or most of the people of a city or a nation, into misery. Moreover, only by maintaining and renewing that state of misery has science been able to make possible its own continued existence. . . . .The most talented and hardworking scientists cling fast to those holding temporal authority, and in exchange for even greater wealth or power or fame, they try to monopolize their own specialized knowledge, and manipulate it, borrowing authority in order to oppress those less fortunate. . . .In short, science and technology are tools of the System, enlarging the power of those who already possess authority, while further weakening and impoverishing those who are ruled by that authority.

If this were true, it would mean that the only role of science and technology was to constantly make the public poor and miserable. But experience shows that this is by no means the case. Figure 8 offers two striking instances to the contrary: it shows that after vaccines were developed for poliomyelitis and Japanese encephalitis, the number of polio and encephalitis patients in Japan declined dramatically, with the

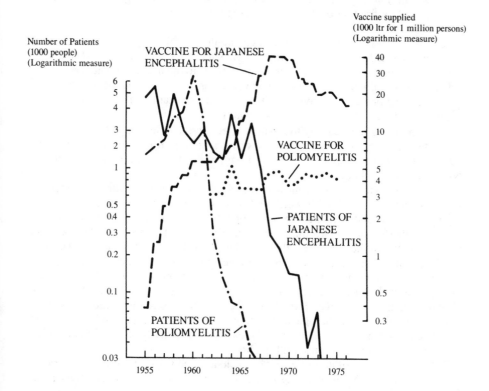

Figure 8   DECREASE OF COMMUNICABLE DISEASES BY THE USE OF VACCINE

increased supply of these vaccines. But Roszak and his followers unfortunately do not seem to notice changes of this kind in their own daily lives. Instead, they seem bent only on stirring up hostility and the fear of science and technology. Some writers have carried this argument to its logical conclusion and advocated the elimination, insofar as possible, of all science and technology. A notable example is the critic Ivan Illich. Born in Vienna in 1926, he studied in Rome and elsewhere, then moved to America in 1951. He is now living in Mexico. Like other advocates of anti-science, Illich believes that modern technology invariably imposes centralized authority, a class structure, and exploitation upon the people. Accordingly, he argues for the elimination of modern science, wherever possible. For example, in discussing transportation, Illich advocates banning all vehicles able to travel faster than a bicycle. A bicycle can go about 15 miles per hour, and there should be restrictions on anything capable of higher speeds. When Illich visited Japan from December, 1980 to January, 1981, he declared during one discussion:

If all means of transportation were restricted to a speed of about 15 miles per hour, one could still reach any point on earth in 16½ days [sic]. If a 15 miles-per-hour speed limit were imposed in most parts of the world, one could still meet with four-fifths of the people on earth within nine days. Besides, the energy needed per passenger mile would be only a fraction of what a jet passenger aircraft uses.

If what Illich says here is correct, then shouldn't he have gone from Mexico to Japan by bicycle? He could have traveled via California, Canada, and Alaska, then crossed the North Pacific on one of those pedalboats found in amusement parks, come ashore at Hokkaido, and switched to a bicycle again to pedal his way down to Tokyo. He claims that one can go anywhere in the world like that in only 16½ days, so he surely would have been able to reach Japan in even less time.

Illich, however, chose to fly to Japan on a jet aircraft. When the other participants in the discussion tried indirectly to question his assertion, Illich retorted: "I came via Narita International Airport. I was fully aware of what I was doing, and I came via Narita. I felt sad about it."

If one believes that ideas can move human society in a particular direction, then the inconsistency between Illich's words and actions cannot be so easily excused just by saying "I felt sad about it." Perhaps intuition told Illich that he could go anywhere in the world in 16½ days at 15 miles per hour. But remember, he arrived in Japan in December. Would he have been able to pedal a bicycle through the ice and snow of Alaska and the Aleutian Islands? Wouldn't we do well to investigate — objectively, elementally, and quantitatively — the question of whether Illich would ever make it to Japan alive? Still, Illich is only criticizing the excesses of technology, not disavowing, altogether, everything technological. He is aware that a modern bicycle is made possible by a technological invention called the ball bearing, and he is pleased that, thanks to that invention, the boundaries of human activity have been expanded. He knows how to compromise with reality.

But the enthusiasts of anti-science also include some who are not willing to make such compromises. Such critics, as could be expected, try to banish all science and technology. They consequently find themselves forced to insist that primitive, uncivilized societies are far superior to today's society. Theodore Roszak, for example, writes:

> The most familiar examples we have of culture dominated by gnosis are in the world's primitive and pagan societies. Many of these societies have been capable of inventing agrarian and hunting technologies every bit as ingenious as the machine technics of modern times. But, in stark contrast to the culture of urban industrialism, their technology blended at every step with poetic insight and the worship of the elements. . . . These "underdeveloped" cultures have proved more technically successful than our own may. They have *endured*, in some cases a hundred times longer than urban industrialism may yet endure.

But today there is no way we can possibly turn back the clock and introduce that kind of primitive society. If anyone in all seriousness advocates a reversion to the "primitive life," he himself ought to be the first to try returning to the kind of existence he advocates, giving up all modern science and technology of any kind. The Japanese soldier mentioned earlier, Shiochi Yokoi, spent 26 years and five months in the jungles of Guam, enduring a primitive existence virtually cut off from modern science and technology. Inside a thicket of bamboo he dug an L-shaped underground den to live in. By pounding the bark of trees, he made fibers which he wove into clothing. He made eel traps by weaving together strips of bamboo, and he burned coconut oil for light at night. Looking back on his 26 years and five months in the jungle, he said of his life there: "I suffered. I really suffered." Someone who sits in a comfortable home, under an electric light, writing manuscripts (published by means of large-scale technology) and insists that primitive societies were better, and that we ought to go back to the primitive life needs to be questioned carefully.

III
THE ENVIRONMENT THREATENED

For twenty years I have watched America grapple with environmental and quality of life issues that are still of major concern. I first visited the United States in May 1964, as a member of a group of journalists touring nuclear power installations in Europe and America. The sites visited by our group included the Dresden Nuclear Power Station (near Chicago) and the Enrico Fermi Nuclear Power Station (where a "breeder" reactor was being tested). Surprisingly, it was not a nuclear facility which gave me my most eye-opening experience; rather, it was the World's Fair that had just opened in New York City.

At the Fair, it seemed I had found the essence of America, a vision of the future, all infinitely brilliant and incredibly wealthy. There were cool shade trees, lush green lawns, bright red tulips, and a multitude of imposing, strangely-shaped, and fascinating buildings. The Chinese Pavilion resembled a dragon-king's palace; glittering gold leaf covered the Hall of Thailand; there were multi-colored overhead cable cars; and hovering, darting sightseeing helicopters. . . .Against the afterglow of the sunset, the lights of the Fair dazzled me with their beauty. Even now I can remember the excitement of the General Electric exhibit, the "Carousel of Progress." I joined the long line of people, waiting an hour to enter. The hall was like a small theater, with perhaps 250 seats. The stage curtain was closed when I heard a voice:

> Now, most carousels just go 'round and 'round without getting anywhere. But on this one, at every turn, we'll be making progress! And progress is not just moving ahead. Progress is dreaming and working and building a better way of life. Progress is a commitment to *people* . . . a commitment to making today and tomorrow "the Best Time of Your Life."

The curtain was still drawn as beautiful, fantasy-like colored lights began to dance across the curtain. The narrator continued:

> It wasn't always easy. At every turn in our history, there was always someone saying, "Turn back! Turn back!" But there is no turning back — not for us, not for our carousel. The challenge always lies ahead. And as long as Man dreams and works and builds together, these years, too, can be "the Best Time of Your Life."

Then a chorus of voices chanted:

Now is the Time —
Now is the Best Time —
Now is the Best Time of Your Life!

Then a sudden silence fell over the 250 seats in the theater as the audience's seats began to revolve slowly about the central stage, like a merry-go-round. When the moving seating area came gently to a halt, I was facing a stage displaying an old-fashioned kitchen and living room. A hand-operated pump stood on the sink in the kitchen, which also featured a real "icebox" (the kind cooled by blocks of ice), a coal-fired cooking stove, and a hand-cranked washing-machine. In an overstuffed armchair sat a gentleman with a small mustache, and a dog lay sleeping on the carpet at his feet. The man spoke:

Well, the robins are back. . . .That's a sure sign of spring. . . .What year is it? Oh, just before the turn of the century. . . .

And believe me, things couldn't be any better than they are today. . . .

Yes sir, we've got all the latest things. . .gas lamps, a telephone and the latest design in cast-iron stoves. . . .

And isn't our new icebox a beauty?. . . .Holds fifty pounds of ice. . . . Milk doesn't sour as quick as it used to. . . .

The dog raised its head, wagged its tail, rolled its eyes, and growled.

Our dog Rover, here, keeps the water in the drip pan from overflowing.

The dog proudly wagged its tail.

This first scene lasted about five minutes, and when it ended, the seats once more began to revolve. A storm of applause rose from the audience. But the dog lying on the stage did not stir, showing not the slightest sign of surprise. That's strange, I exclaimed to myself. Only then did I realize that "well-trained" Rover was a robot. From my moving seat, I strained to look at the actors in the scenes that followed. Were they real? The actors moved their eyebrows, furrowed their foreheads, turned their eyes and bent their fingers. Crossing his legs, one actor tapped his foot in time with the musical accompaniment. But looking more carefully, I felt that they were not humans. It would have been very hard for me to point out exactly what made them different from actual people, but there seemed to be something artificial about them. "They're artificial. . .I'm sure of it," I said to reassure myself.

These manmade players, I later learned, were the new products of a technique called "Audio-Animatronics," developed by Walt Disney Enterprises after more than ten years of research, at a cost of over a million dollars. Beneath their flexible, lifelike plastic skins, the figures were equipped with a number of air- or water-filled bags or sacs. By increasing or decreasing the air or water supply to these sacs, the expansion and contraction of human muscles could be imitated. As more water, say, was forced into a sac beneath the facial skin, the wrinkles or lines on that "actor's" face would vanish; and as the sac was emptied, the wrinkles would reappear. Similarly, as a sac expanded, it would straighten a character's finger into a pointing gesture, and relax the finger as it emptied. As I later learned from a public relations representative at the General Electric exhibit, the G.E. Carousel of Progress was "peopled" by a total of 32 robots.

The second scene of the performance showed an American household of the 1920's. The husband and father spoke to the audience:

Well, we've progressed a long way since the turn of the century twenty years ago. . . .But no one realized then that this would be the Age of Electricity. Everyone's using it! Farmers, factories, whole towns. . . .

He pointed to all of the electrical appliances in his own home:

Electric sewing machine. . .coffee percolator. . . toaster. . .waffle iron. . .refrigerator. And they all go to work at the click of a switch!

The phonograph was in everyday use, and radios were no longer a novelty. On the stage, a radio news program reported that Charles Lindbergh had successfully flown solo across the Atlantic Ocean. The third scene was set in the 1940's, while the fourth and final scene showed what the 1970's would bring.

Each time a scene concluded, the robots chanted:

Now is the time —
Now is the Best Time!

The world's forward marching
And you're in the parade!

Now is the time —
Now is the Best Time!

As one era passed from the stage, that "best time" gave way to the even better era that followed. The overall theme was that the future would be even better than the present.

I also visited the panoramic theater sponsored by General Motors, call "Futurama." As soon as I entered, a seat came gliding in smoothly toward me. When I sat down, I was carried into a pitch-black tunnel, a "time-tunnel," beyond which lay the world of the future. There, lunar rovers ran about on the surface of the moon, while in an undersea vacation resort, the guests at a luxurious sea-bottom hotel were enjoying the wonders of the underwater world. Giant jungle trees were being felled by laser beams, and huge construction vehicles built and paved highways before my eyes. Here, too, was the vision of an ever brighter future. And everything at the World's Fair made it clear that science and technology would be responsible for that bright future.

Ashes of Death

General Motors had presented Futurama earlier, at the 1939-40 New York World's Fair, held on the exact same site. That Futurama had also been a tremendous success. In 1939, the Futurama had depicted television, streamlined automobiles, and electrical living, and General Motors was proud that most of the future predicted then had become a reality by 1964. "Practically everything predicted in this Futurama will probably come true, too," said the guides at the exhibit. But in 1964 the germs of anti-science thinking had already slipped into America, and were slowly and steadily multiplying. One reason for the growth of anti-science was the series of nuclear weapons tests that lasted from the 1950's into the early 1960's. The atomic bombs that fell on Hiroshima and Nagasaki exploded with the force equivalence of about 20,000 tons of the high explosive TNT. The conventional "blockbuster" bombs used in the second World War carried 20 tons of high explosive. When the war ended, the atomic bomb seemed America's treasured secret. But in only four years the Soviet Union caught up with the United States. President Truman made the announcement himself on September 23, 1949: "We have evidence that within recent weeks an atomic explosion occurred within the USSR." Two days later, the Soviet news agency, Tass, confirmed the report. At a White House press conference about two weeks later (October 6) a reporter asked:

Mr. President, do you think it is possible, in view of the atomic thing, to reach an agreement with Russia any more than it was before the atomic —

President Truman cut in:

I can't answer a question like that. I don't know. We have made the most important proposition in the history of the world with regard to that atomic situation, and the Russians didn't see fit to accept our proposition.

Then, on January 31, 1950, President Truman instructed the Atomic Energy Commission to begin development of a hydrogen bomb. That order heralded the beginning of a dark era.

The "ashes of death" drifted down upon the world. On February 26, 1952, Prime Minister Churchill announced that Britain also had the atomic bomb and held their first nuclear test on October 3. Shortly thereafter, on November 16, the United States announced the first successful test of a hydrogen bomb on the Eniwetok atoll in the Pacific.

The Soviet Union tested its first hydrogen weapon in August, 1953. From then on, America and the Soviet Union engaged in a desperate competition to develop more and better thermonuclear weapons. In the midst of this nuclear race, a Japanese fishing vessel, the *Fukuryu Maru No. 5* (also called the *Lucky Dragon*) was showered with nuclear fallout, the "ashes of death," near the Bikini atoll in the Pacific. This occurred on March 1, 1954. The crew was catching tuna when they saw, on the horizon, an inverted triangular cloud. Five or six minutes later, they heard a tremendous blast, like half a dozen simultaneous thunder claps. About an hour later, soft white ash began falling upon the boat.

The 23 fishermen found that before long their exposed skin turned dark, became inflamed and blistered. The ship returned to the harbor of Yaizu, in Shizuoka Prefecture, Japan, on March 14. The crew members were hospitalized immediately. One fisherman, Aikichi Kuboyama, died on September 23.

About that time, "ashes of death" produced by American and Soviet H-bomb experiments were falling upon a number of localities in Japan. Rain became radioactive. Figure 9 shows in graphic form the amount of radioactive fallout on Japan.

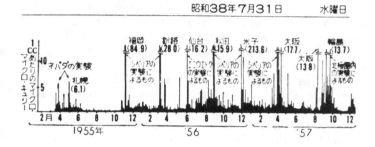

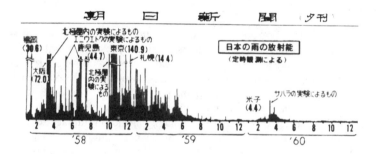

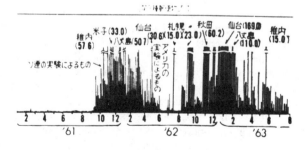

This diagram shows the measurement of radioactive rain resulting from U.S. and Soviet H-bomb experiments from the period 1955 to 1963.

Figure 9   FALLOUT IN JAPANESE RAIN

This graph makes clear how heavy the fallout was and for how long it continued. A group of housewives in Tokyo, unable to endure the situation any longer, started a movement opposing nuclear weapons testing. The movement spread worldwide and in October, 1963, the partial nuclear test-ban treaty went into effect, putting an end to American, British, and Soviet nuclear tests in the atmosphere.

Many people witnessing the competition to develop thermonuclear weapons had begun to feel uneasy about the "runaway" implications of science and technology. On October 4, 1957, when the Soviet Union launched *Sputnik 1*, the world's first artificial satellite, a great outcry went up from the American people: Don't let the Russians beat us! The rockets used to launch *Sputnik* could obviously also carry atomic and thermonuclear weapons. People worried about the "missile gap" between the United States and the USSR; they called for massive funding to meet the challenge and to develop science and technology.

Large sums of money were spent to bring outstanding foreign scientists to America, initiating the "brain drain." American scientists and engineers, at the same time, were able to visit laboratories and factories abroad. At this juncture, President Eisenhower cast the first stone in the direction of science and technology. On January 17, 1961, he delivered his Presidential farewell address on radio and television. In this speech, he warned "my fellow Americans" about too close a relationship between America's military and industry:

> Until the latest of our world conflicts, the United States had no arma-
> ments industry. American makers of plowshares could, with time and as
> required, make swords as well. But now we can no longer risk emer-
> gency improvisation of national defense; we have been compelled to
> create a permanent armaments industry of vast proportions. Added to
> this, three and a half million men and women are directly engaged in
> the defense establishment. We annually spend on military security more
> than the net income of all United States corporations. . . .This con-
> junction of an immense military establishment and a large arms industry
> is new in the American experience.

After pointing out that this "military-industrial complex" was exerting its influence upon federal, state, and local governments, he said, "We must never let the weight of this combination endanger our liberties or democratic processes." He then turned to the "technological revolution." He pointed out that the sudden change in the relationship between industry and the military was due to a technological revolution over the preceding several decades. As a result of this revolution, "research" had been placed at the center of everything. Research was no longer the lonely individual's pursuit it had once been, but had turned into a task performed by large groups of scientists in big laboratories. As research had become more and more expensive, the moving force that spurred research was no longer intellectual curiosity, but government funding. "For every old blackboard," President Eisenhower declared, "there are now hundreds of new electronic computers." The President warned:

> Yet, in holding scientific research and discovery in respect, as we
> should, we must also be alert to the equal and opposite danger that pub-
> lic policy could itself become the captive of a scientific-technological
> elite.

The day after President Eisenhower's farewell address, his scientific adviser, Dr. George B. Kistiakowski, received telephone calls from a number of scientists, asking him whether the President had turned anti-scientific. That evening, at a farewell party for the members of the White House staff, President Eisenhower asked Kistiakowski about the reaction to his farewell address. When Kistiakowski mentioned all the telephone calls he had received, the President appeared extremely surprised. After the party, the President invited Kistiakowski into his private office, ordered drinks, and began explaining his position. The President explained that he unhesitatingly supported basic academic research, and that he feared only the rising power of a militarily-oriented science. He asked Kistiakowski to try to make this distinction clear to the public.

This speech of President Eisenhower is strangely similar in tone to some of the concerns expressed by anti-science advocates: "science and technology are tools of the Establishment," or "science and technology are servants of War." The source of Eisenhower's concern may have been simply the possibility that the military-industrial complex would grow too large; but without intending to do so, the President stimulated the germs of anti-science attitudes.

Silent Chemicals

As the American public was starting to feel vaguely uneasy about science and technology a major shock came with the publication of Rachel Carson's *Silent Spring*, a work which warned that improper and excessive use of insecticides would destroy the ecosystem. Rachel Carson was born in a small town in Pennsylvania. There was a dense woods, full of butterflies and birds, where she had played as a child. When she entered Pennsylvania State College for Women, she intended to become a writer, but during her second year, she became fascinated with biology. Later she studied at Johns Hopkins University and the Marine Biological Laboratory at Woods Hole, Massachusetts. She taught at Johns Hopkins and the University of Maryland, and then went to work for the U.S. Fish and Wildlife Service. Her interests in biology and writing gradually came together.

In 1941, she wrote a book called *Under the Sea Wind*, just one week before Japan attacked Pearl Harbor. Americans, swept into the maelstrom of the second World War, had no time to read a lyrical work about the sea. The book drew little attention, but *The Sea Around Us*, published in 1951, was a tremendous success, appearing on the best-seller list for 86 weeks. Rachel Carson had suddenly became a leading writer. A few years later (January, 1958) she received a letter from a friend, Olga Owens Huckins, lamenting that DDT sprayed from aircraft for mosquito eradication had resulted in the deaths of numerous songbirds.

Reading the letter reminded her that this problem had been on her mind for some time, and she decided, then and there, as she later said, that "I must write a book." That book was *Silent Spring*. In the course of writing her books on the sea, she had befriended a number of marine biologists. She learned from them that DDT could move up through the aquatic "food chain" and accumulate in the flesh of fish. It had already been reported, in 1957, that a DDT-contaminated food chain had been discovered in Clear Lake, in California. According to the report, the water of the lake contained DDD (an insecticide very similar to DDT) at a concentration of 0.02 parts per million (ppm). But in the bodies of microscopic plankton, the concentration of

DDD rose to 5 ppm. And in the bodies of fish that normally ate large quantities of that plankton, the amount of DDD was 2,000 ppm. DDT and DDD easily accumulate in fatty tissue. The fish, in turn, were eaten by grebes, a type of diving bird. And the grebes were dying in large numbers. Rachel Carson compiled large quantities of scientific data on this, gathered research, and wrote *Silent Spring*:

> Then a strange blight crept over the area and everything began to change. Some evil spell had settled on the community: mysterious maladies swept the flocks of chickens; the cattle and sheep sickened and died. Everywhere was a shadow of death. The farmers spoke of much illness among their families. In the town the doctors had become more and more puzzled by new kinds of sickness appearing among their patients. There had been several sudden and unexplained deaths, not only among adults but even among children, who would be stricken suddenly while at play and die within a few hours.
>
> There was a strange stillness. The birds, for example — where had they gone? Many people spoke of them, puzzled and disturbed. The feeding stations in the backyards were deserted. The few birds seen anywhere were moribund; they trembled violently and could not fly. It was a spring without voices.
>
> This town does not actually exist, but it might easily have a thousand counterparts in America or elsewhere in the world.

Her book was a frightening narrative with pages filled with words like "dying," "killing," "slaughter," "massacre," and "extermination." *Silent Spring* went on sale in September, 1962, and by December, 100,000 copies had been sold. By the end of the year, at least forty states had approved legislation dealing with insecticides, and the President's Science Advisory Committee issued a report on the subject on May 15, 1963. When the report was released, a CBS news commentator, Eric Sevareid, noted:

> Miss Carson had two immediate aims. One was to alert the public; the second, to build a fire under the Government. She accomplished the first months ago. Tonight's report by the Presidential panel is *prima facie* evidence that she has also accomplished the second.

Thanks to her literary genius, millions of Americans were now clearly aware of the horrors of insecticides. But in order to do this she deliberately omitted from her book one very important fact: namely, the significant benefits people derive from the use of insecticides. DDT was first used on a large scale in Italy during the second World War (1943-44) when an epidemic of eruptive typhus broke out in Naples. The symptoms of typhus (which is carried by body lice) include chills, fever, headaches, and pain in the joints. Four days after infection, red blotches about one-eighth of an inch in diameter break out all over the body. On about the seventh day, the victim becomes delirious and frenzied. In the past, about 30 percent of those stricken with typhus died.

In Naples, the American armed forces spread large quantities of DDT to kill body lice, and succeeded in stopping the typhus epidemic. Before DDT appeared on the scene, fleas, lice, ticks, mosquitoes, and flies were common in almost every country. The extent to which insecticides have improved our environmental sanitation and reduced disease is almost incalculable. DDT, in fact, used to be called "the atomic bomb of insecticides." Rachel Carson, however, made no mention of the benefits from insecticides. Although nowhere in *Silent Spring* did the author advocate a total ban on insecticides, a hasty reader might form the impression that all use of insecticides should be prohibited. Moreover, when the public learned about the harm that insecticides did, it began to feel vaguely uneasy about science and technology, since insecticides had been developed through the efforts of science and technology.

In 1962, when *Silent Spring* was published, the public was equally shocked by another frightening report that revealed birth defects caused by the drug thalidomide. The first American publication to report extensively on this episode was the weekly magazine *Time*. Its issue for February 23, 1962, carried an article headed "Sleeping Pill Nightmare," reporting on the link between a birth deformity known as phocomelia (characterized by flipper-like arms and legs) and the sedative thalidomide. The article, which was accompanied by photographs of deformed fetuses, shocked the American public. About two months later (April 11), Dr. Helen B. Tausig, Professor of Pediatrics at Johns Hopkins University Hospital, described the phocomelia deformity to the annual meeting of the American College of Surgeons. The *New York Times* carried her report the next day, under a headline stating that this birth defect was caused by a drug.

Thalidomide was a sedative that had been invented in West Germany, and sold there and in Britain since 1958. Since thalidomide users woke up the next morning feeling refreshed and alert, it was for a time considered the best of all sleeping drugs. It could also be used to relieve the "morning sickness" of early pregnancy, so it was frequently prescribed by physicians in West Germany, Britain, and Japan for women with severe morning sickness. In February, 1961, however, Dr. Widukind Lenz of West Germany revealed his suspicions that thalidomide caused limb deformities in fetuses. If a woman took this drug early in her pregnancy, her baby might be born without the long bones of the arms and legs, and with hands and feet attached directly to the body. From the resulting resemblance to the flippers of a seal, this deformity is called phocomelia, from *phoca*, the Latin word for seal. It has been estimated that 3,500 to 5,000 such malformed children were born in West Germany, and 300 to 500 in Britain.

In America, in September 1960, the William S. Merrell Company applied to the Food and Drug Administration (FDA) for a license to market thalidomide. But Dr. Frances O. Kelsey of the FDA had doubts about the drug's safety, and refused to allow a license to be granted. Because of her prudence, thalidomide was used in America only on an experimental basis. In Japan, the Dainippon Pharmaceutical Company had been selling the sedative, but it halted all shipments on May 17, 1962. In *Silent Spring*, Rachel Carson had vividly described the risk of hereditary defects, such as malformed children, caused by insecticides. Thus, in the minds of many laymen, there was a kind of overlapping between insecticides and thalidomide, and this tended to reinforce fears of science and technology.

The harm done by herbicides was also described by Rachel Carson in *Silent Spring*:

> The legend that the herbicides are toxic only to plants and so pose no threat to animal life has been widely disseminated, but unfortunately it

is not true. The plant killers include a large variety of chemicals that act on animal tissue as well as on vegetation. They vary greatly in their action on the organism. Some are general poisons, some are powerful stimulants of metabolism, causing a fatal rise in body temperature, some induce malignant tumors either alone or in partnership with other chemicals, some strike at the genetic material of the race by causing gene mutations.

The United States began spraying herbicides on the jungles of Vietnam in 1962, in what was referred to as a "defoliation campaign." The idea was to strip away the covering foliage and expose the movements of guerrillas concealed in the jungle. The herbicides were spread in large quantities from military aircraft. The area sprayed increased year by year: only nine square miles in 1962, but 343 square miles in 1965, and 2,668 square miles in 1967. In the middle of 1962, 12,000 American combat soldiers were sent to Vietnam, increasing to 23,000 by early 1965 and then to 181,000 at the end of that year. Finally, the number of U.S. troops rose to 486,000 by the end of 1967. As American involvement in the Vietnam conflict escalated, opposition to the war developed at home. In March, 1964, 6,000 people took part in demonstrations in New York City, demanding that the "aggression in Vietnam" be halted. On March 31, 1965, 300 prominent Americans signed an open letter to President Johnson, printed in the *New York Times*, calling for "an immediate cease-fire in Vietnam." The letter declared that "the bombings of North Vietnam and the landing of the Marines in South Vietnam constitute an open invitation to world war." The number and size of anti-Vietnam demonstrations increased. Those who opposed the war also objected to the direct utilization of science and technology in the conduct of the war: the defoliation campaign was a prime example of this.

On November 30, 1965, Ralph Nader's *Unsafe at Any Speed* appeared, a vehement and emotional attack on the safety of American automobiles and on their manufacturers. The theme of the book was this: American car manufacturers are well aware that their products are full of defects and extremely dangerous, but in their greed for profits, the companies sell cars without taking steps to improve their safety.

Ralph Nader, born in Winsted, Connecticut in 1934, received degrees from Princeton University and Harvard University Law School. After six months of military service, he opened a legal practice in 1959. For three years, however, beginning in 1961, he worked as a free-lance journalist, traveling to Europe, the Soviet Union, Africa, and South America. In 1964 he became a personal assistant to Daniel P. Moynihan, then an Assistant Secretary of Labor, where he investigated the safety of expressways and automobiles. The results of his research became the basis for the book *Unsafe at Any Speed*. The book attracted hardly any attention; the publisher, anxious to raise sales, visited one newspaper office after another, asking them to review the book. He even arranged a press conference in Detroit, the home of automobiles. As a result, the initial press run of 9,000 copies was soon exhausted and by February, 1966, the book had sold 22,000 copies. But still it did not make the headlines.

Then, however, General Motors decided to hire a private investigator and inquire into Nader's private life. They were motivated to do so by the very first chapter of Nader's book, headed "The Sporty Corvair," which attacked GM's compact Corvair. According to Nader, faulty design in the Corvair made it extremely prone to skid and overturn at high speeds.

The hired investigators, Vincent Gillen Associates, were told to find out why Nader was interested in automobile safety, who was helping him, what his personal ties to women were, whether he used drugs, etc. One private investigator was even assigned to "tail" Nader. Nader noticed he was being followed and soon found out that GM had hired the investigators. The whole episode was reported in the *Washington Post* and the *New Republic* during March, 1966. A U.S. Senate subcommittee summoned the head of GM to testify about the incident. The revelations about GM's attempts to investigate Ralph Nader turned him almost overnight into a popular hero. *Unsafe at Any Speed* became an immediate best seller, with the hard-cover and paperback editions selling, respectively, 70,000 and 250,000 copies.

The automobile had been thought of as the symbol of social progress, and the growth of the automobile industry seemed the symbolic model of America's economic development. But Nader's book made the automobile a "public enemy," and made the Detroit automakers seem like "murderers." Technicians, in general, were viewed as accomplices in this "murder." Nader wrote: "They did not have the professional stamina to defend their engineering principles from the predatory clutches of the cost-cutters and stylists."

Other obvious symptoms of this anti-science attitude began to appear about this time. In December, 1965, I returned to the United States to participate in some of the actual training of American astronauts. At Wright-Patterson Air Force Base, near Dayton, Ohio, I rode in a special jet aircraft called "The Wonder of Weightlessness," which created the experience of the weightless conditions of space flight. At the Aerospace Medical School in San Antonio, Texas, I whirled around in a centrifuge and wore a complete space suit. The experience was exhilarating. The development of space was a magnificent vision of mankind. But something had changed. The America I saw at the end of 1965 was different. Many people had begun criticizing the space program with comments such as the following:

> Spending money on trips to the moon or to Mars is like pouring money down a rathole. There's nothing particularly fascinating about Mars or the moon. They must just be uninhabited wastelands. If there's money to spend on going to places like that, I'd like them to repair the street in front of my house. They ought to build apartments for black people. It's more important to get rid of poverty and cure disease here on earth than to travel around in space.
>
> Suppose they find gold on the moon — it'll never be profitable, because it would cost too much to bring back to earth. Anything that doesn't pay for itself should be stopped.
>
> Those huge buildings at the Kennedy Space Center will probably be the laughingstock of future generations, just as we look at the pyramids and wonder how the ancient Egyptians could have built anything so foolish and useless.

Everywhere I went I heard people saying things like this. Then, when I visited the Manned Space Flight Center (now the Johnson Space Center) in Houston, I met Paul Pauser, then special assistant to the director of the center. Before I even brought up the subject, he spoke with enthusiasm about America's reasons for exploring space:

Up to now, so much of our progress in science and technology has occurred during wartime. In order to win, we have poured huge amounts of national wealth and scientific manpower into weapons research. Because of those efforts, radar was invented, the jet aircraft was developed, the atom was harnessed. To allow science to progress in peacetime at the same pace as in wartime, we need large-scale programs like the Apollo project. The American people want this country to beat the Soviet Union in exploring and developing space. Even if we spent huge amounts of money for better medical treatment and radios and refrigerators to raise our standard of living here on earth, the American people wouldn't be at all happy to see the Russians defeat us in the race for space. The moon project fulfills that particular strong demand of the American people, but at the same time, it will stimulate technological progress that will help to improve life back here on earth.

Project Nohole

Even in 1984 the statements from two decades ago have remarkable currency. The editor of the prestigious magazine *Science*, Philip H. Abelson, wrote the following in an editorial entitled "National Science Policy," for the January 28, 1966, issue:

Federally supported research activities are being reexamined. The immediate cause is the budgetary squeeze brought on by the war in Vietnam. More fundamental is the fact that a 20-year honeymoon for science is drawing to a close. Although needs for support of basic research are increasing, expanded budgets will be obtained only after convincing justification has been provided. Indications of the present climate can be seen in the executive branch, the press, and Congress. In contrast to other years, President Johnson hardly mentioned science in his State of the Union speech.

According to Abelson, science enjoyed a "honeymoon" after the end of the second World War. Some people have called the era the "golden age" of science in America. Particularly after the Soviet Union launched *Sputnik 1* on October 4, 1957, the American scientific community had every reason to rejoice in its good fortune. At the time, the United States was confronting the Soviet Union as the "cold war" unfolded. In the midst of this rivalry, America found itself beaten to the punch by the Soviets, with the historic launching of the first artificial satellite. The echoes were everywhere: "Are America's science and technology inferior to the Russians'? . . . Unless we do something, the technology gap is going to keep widening."

Then, in a series of space exploits, the Soviet Union became the first to impact a rocket vehicle on the moon's surface; the first to photograph the hidden side of the moon; and the first to send a man aloft in orbit. The conviction in the U.S. that something had to be done grew stronger. When John F. Kennedy became President, in January, 1961, he turned out to be a genuine "guardian angel" for the promotion of science and technology. On May 25, just a short time after taking office, President

Kennedy announced, as a national goal, a project to land men on the moon — the Apollo program. The President addressed Congress with these words:

> First, I believe that this nation should commit itself to achieving the goal, before this decade is out, of landing a man on the moon and returning him safely to the earth. No single space project in this period will be more impressive to mankind, or more important for the long-range exploration of space; and none will be so difficult or expensive to accomplish. . . .I believe we should go to the moon.

On September 12, 1962, in a speech delivered at Rice University in Houston, President Kennedy said:

> We choose to go to the moon. We choose to go to the moon in this decade and do the other things, not because they are easy, but because they are hard, because that goal will serve to organize and measure the best of our energies and skills, because that challenge is one that we are willing to accept, one we are unwilling to postpone, and one which we intend to win, and the others, too.
>
> To be sure, we are behind, and will be behind for some time in manned flight. But we do not intend to stay behind, and in this decade we shall make up and move ahead.

Kennedy increased the budget for space development, and gave his full backing to the scientists and engineers. A Japanese biochemist, Dr. Sen'ichiro Hakomori, who was invited to America in 1963 and later decided to stay, reflected on those times:

> As I look back on it now, there was a kind of "Kennedy boom" period, when America's research spending increased sharply. Scientists who had barely been able to get funding before then suddenly found research money pouring in. They didn't know what to do. They wrote letters all over, even to people they hardly knew, hoping they could grab somebody. That's when I got the invitation from Professor Jeanloz (of Harvard University) and came to the United States. But I was disappointed to find that nothing particularly exciting was going on. Afterwards, though, I decided to stay.
>
> In those days, America had the money, I guess, and the only problem was how to spend it. If you just sent in an application with a brief review of your work, you could get the funding, and that was when the dollar was still strong, too.

On November 22, 1963, President Kennedy, the generous patron of the scientists and engineers, was assassinated in Dallas. His successor, Lyndon Johnson, was not so favorably disposed toward science and technology. It was January, 1966, when *Science* editor, Abelson, wrote that the "honeymoon for science is drawing to a close," and on June 27 of that year, President Johnson summoned a group of leaders of the medical community to discuss spending for research. After this meeting, the President said:

The National Institutes of Health are spending more than 800 million dollars a year on biomedical research. I am keenly interested to learn not only what knowledge this buys but what are the payoffs in terms of healthy lives for our citizens.

Noting those comments by the President, John Fischer, the editor of *Harper's* magazine, said that this brought down the curtain on the era when science reigned supreme in the government's eyes. In fact, in August of 1966 Project Mohole was cancelled. This program, initiated in 1960, was to drill a deep shaft into the ocean floor northeast of Hawaii, to study the interior of the earth.

When the program began, it was budgeted for 47 million dollars; but construction of the necessary seaborne platform alone cost 79.6 million dollars, and it became clear that even after the actual drilling had begun, costs would still run to 11 million dollars annually for at least three years. The U.S. House of Representatives, by a vote of 108 to 59, removed all funding for the program for the fiscal year 1967 budget. Project Mohole ended up as "Project Nohole." *Science's* editor Philip Abelson, extremely concerned about the situation, wrote:

The United States enjoys world leadership and is confident of its role. More precisely, we are overconfident. The spirit of urgency that followed Sputnik has evaporated. A mood of relaxation has taken over that perhaps will not be broken until again we feel mortally threatened.

The spirit of the times is manifested in many ways; one of them is in the support of scientific research. The nation is willing to gamble future scientific leadership in order to "save" a few hundred million dollars. Every field of science is feeling the consequences of budgetary tightening. Many competent scientists cannot obtain support for their research. The morale of the scientific community has been damaged.

A classic example of this situation, wrote Abelson, was the fate of Project Mohole. He noted that Mohole was an extremely important program for learning about the structure of our planet and investigating natural resources beneath the ocean floor. He broadened his argument:

We have been leaders in fundamental research. We have been able to attract some of the best talent from abroad. We were able to support many of the world's top scientists in their homelands. Associated with academic research has been the training of industry's and government's scientific talent. A by-product of the grants system has been development of a great instrumentation industry helpful to all segments of science, including industry, and beneficial to our balance of payments. Coupled with our competence in science has been technological strength, which is at the core of financial and military strength.

Abelson goes on to say, under the heading "Penny Wise, Pound Foolish": Americans failed to take this question seriously. Perhaps it was because too much money had been spent on the war in Vietnam. As I have mentioned earlier, there is little doubt that an anti-science and anti-technology mood, in part aroused by books like *Silent Spring* and *Unsafe at Any Speed*, had a major influence on the political decisions to cut funding for science. Clearly, America was coming down with the anti-science disease. It was during the years 1965 and 1966 that the symptoms of the disease became obvious.

Japan, on the other hand, was still immune to the disease. That is evident from Figures 10 and 11. Figure 10 shows the number of copies of *Silent Spring* printed in Japan every year from 1964 to 1980. The Japanese edition, entitled *Elixir of Life and Death*, was first published in 1964 by Shinchosha, Ltd., but attracted little attention. Because the examples in the book were so American, Japanese could not identify with the problem. Agriculture in Japan is on a much smaller scale than in America, with virtually none of the immense expanses of acreage that are so common in the United States. Large-scale spraying of insecticides from aircraft, for example, is a rarity in Japan. Thus, the great majority of readers in Japan probably did not view the concerns of *Silent Spring* as applicable to Japan.

In America, within three months after its publication, 100,000 copies of *Silent Spring* had been sold; but in Japan, the initial printing was 7,000 copies, while the copies printed in each of the following two years numbered only 5,000. In 1968 and 1969, 2,500 copies were printed. In 1970, however, sales of the book suddenly shot upward, and 27,500 copies were printed.

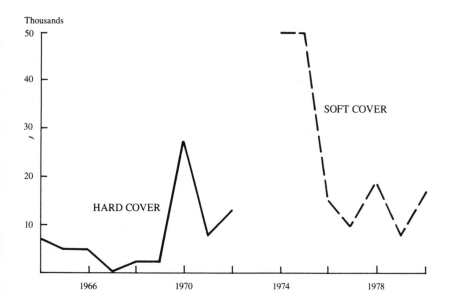

Figure 10   COPIES OF *SILENT SPRING* PRINTED IN JAPAN

56

Figure 11 shows how Japanese attitudes toward nature changed between 1953 and 1978. The graph is based on a study, the *Study of Japanese National Character*, conducted by the Institute of Statistical Mathematics of the Japanese Ministry of Education. The survey includes this question:
Of the following opinions about the relationship between Nature and mankind, which one do you consider closest to the truth?

1. For mankind to be happy, we must yield to Nature.
2. For mankind to be happy, we must utilize Nature.
3. For mankind to be happy, we must conquer Nature.
4. Other than above (please state).
5. No answer.

The graph reveals that in 1968 respondents with an activist attitude (choosing "conquer Nature") comprised 34 percent of the entire sampling. By contrast, those showing a passive viewpoint (choosing "yield to Nature") were only 19 percent of the total. In the 1973 survey, however, those two positions were reversed. People who favored "yielding to Nature" made up 31 percent of the whole group, while those wishing to "conquer Nature" had dwindled to 17 percent.

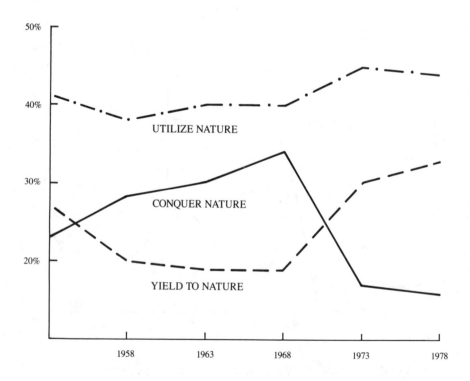

Figure 11   JAPANESE ATTITUDES TOWARD NATURE

IV

THE ESTABLISHMENT UNDER ATTACK

Unrest on college campuses began in America in the autumn of 1964, sparked by the "Free Speech Movement" at the University of California, Berkeley. When University authorities tried to ban on-campus political activities by Berkeley students, about a hundred graduate students and teaching assistants marched in a demonstration on November 10, 1964. More demonstrations followed, and the confrontation escalated until, on December 2, a thousand students occupied the University's administration building. On December 3, the police were called in, and arrested 796 students. This mass arrest ignited a conflict that was to continue.

The Berkeley movement set the tone for student unrest at campuses all over the free world during the latter half of the 1960's. According to those who have studied the reasons for this, the first of its several causes was the emergence of large-scale universities, or "megaversities," with sizeable graduate schools. The number of university students rose sharply after the second World War. In America, where only one million attended college in 1930, there were 1,675,000 college students in 1945, and by 1970 the total enrollment had reached seven million. To manage these huge student populations, university administrators turned to computerization for enrollment, billing, and registration procedures. Even the grading of examinations was done by computers. Television sets were used in large classes. During the turmoil at Berkeley, one of the favorite signs carried by the demonstrators read: "I Am a Human Being: Please Do Not Fold, Spindle, or Mutilate."

In the "megaversities," many of the lectures were delivered over public address systems in huge auditoriums, depriving students of all personal contact with the instructor. Teaching assistants substituted for professors. Many students felt they had been taught only fragmentary bits of scientific research. They felt their needs unfulfilled by assembly-line instruction and an impersonal, computerized campus environment. A few students gave voice to these dissatisfactions, perhaps no more than five percent of the total student body. In his book *Rebellion in the University*, Seymour M. Lipset notes that in America, "five percent of the student body equals over three hundred and fifty thousand students."

Following the Berkeley uprising, campus struggles became more violent. If one reads the statements by the leaders of the anti-university struggle of those days, one finds assertions like this: "The managerial society has been shaped by the development of science, and we are in danger of being treated as 'things.' Our rejection of that managerial society has become the driving force behind our movement." Or: "The objective of the university struggle is to destroy the places where science as it exists has found sanctuary, to destroy its very manner of existence." Because universities have played an important role in scientific and technological progress, the students who led the campus uprisings aimed part of their attack at this particular aspect of the universities.

At the heart of the student movement were concepts like "Establishment" (or "System") and "anti-Establishment," thinking that divided society into "those who rule" and "those who are ruled." Karl Marx's nineteenth-century distinction between the capitalist class and the laboring class has been a favorite distinction of left-wing activists for over a hundred years. Two decades after the end of the second World War, however, large numbers of people in the advanced industrialized countries had become prosperous enough to think of themselves as belonging to the middle class. With this, the Marxist distinction between "capitalists" and "workers" began to lose some of its power as a slogan for social conflict.

Concepts that emerged in response to this development were Establishment and anti-Establishment. Those favoring these terms were able to lump most of the population into one grouping — those who are ruled or controlled — regardless of whether or not they owned stock, or whether or not they belonged to the middle class. Moreover, it was taken for granted that the System or Establishment was always evil, without exception. Accordingly, university authorities are always seen as part of the "Establishment" or "System" that must be overthrown. Newspapers and magazines, intellectuals and critics began to adopt this view of society. For the latter half of the 1960's, the terms "Establishment" and "anti-Establishment" became everyday catchwords.

"Man is an animal that needs a master," said the great German philosopher Immanuel Kant, in an accurate observation. Man is basically a creature that lives in groups, and frequently engages in mass activities. Such activities, unless firm leadership is present, are likely to end in confusion and failure. Leaders or coordinators play an extremely important role. If a leader makes an error in judgement or a wrong decision, the entire group can be inconvenienced, suffer harm, or go off course. The problem is how to choose good leaders who are an asset to the group, and how to remove leaders who are detrimental to the interests of the group.

In a totalitarian state, leaders are usually selected through violent struggle, very similar to the process by which a troop of monkeys picks its "boss." The male monkeys bare their teeth and fight, with the winner as the leader. In a democratic country, government leaders are chosen through election, resulting in a system whereby the "ruled" select the "rulers." Of course elections can sometimes be corrupt, or not truly free. Even in democratic countries, private businesses and universities are not totally democratic. In most corporations, the chairman and the board of directors are selected by stockholders' votes. An employee of the firm, unless he or she owns stock in the company, has no right to participate in the voting process. In most universities, the president is ultimately chosen by a vote of the trustees, not by the students. Thus, it would be accurate to say that in many or most corporations and universities, the "ruled" do not get to choose their "rulers." In a democratic country, however, to restrain official high-handedness and oppression, it is legally permissible to form groups like labor unions and student self-government associations.

Most of the so-called "progressives" or left-wing intellectuals of the free-world countries uncritically accepted the logic of the anti-Establishment. The result was that the anti-Establishment movement swept through the countries of the free world like a great whirlwind. Out of this turmoil emerged the notion that science and technology were "tools of the System." Although science and technology actually benefit the general public, not just the leadership class (see Figure 8 in Chapter II), the overall benefits of science and technology were virtually ignored by the anti-Establishment zealots. The notion that they are nothing but "tools of the System" gained wide currency as anti-Establishment agitation grew more intense. In the arguments of the anti-Establishment movement, even Marxism came in for sharp criticism, on the grounds that Karl Marx himself had a high regard for science and technology. Marxism is sometimes referred to as "scientific" socialism, in contrast to the "utopian" socialism that preceded it, but the "scientism" of Marxism was nevertheless denounced by the anti-Establishment people. Most of the leaders of the anti-Establishment movement held anti-scientific and anti-technological views. As one might expect, during the late 1960's and on into the 1970's, when the anti-Establishment movement was at its strongest, the anti-science disease was at its worst.

The Wrong Stuff

In the climate just described, it was science and technology, ironically, that further aggravated the anti-science disease. An example of this was the historic first orbiting of the moon by a manned spacecraft, achieved by *Apollo 8* on December 24, 1968. Frank Borman and two other American astronauts were aboard. At the time, the United States was still engaged in competition with the Soviet Union to explore and exploit space. International attention was focused on the outcome: which country would be first to land a man on the moon? America had just completed the *Apollo 8* spacecraft in October that year. The one before it, *Apollo 7*, had orbited the earth for eleven days (October 11-22) carrying Walter M. Schirra as commander. Originally, *Apollo 8* was supposed to orbit the earth to check out the safety of the space vehicle, and then the next spaceship was to head towards the moon. There was concern, however, that by spending that length of time in an earth orbit, the Americans might be giving the Soviets a chance to beat them to a lunar landing.

At this point, NASA decided to skip several of the intermediate steps, and ordered *Apollo 8* to go for lunar orbit. Dr. Bernard Lovell, director of Britain's Jodrell Bank Radio-Astronomical Observatory, called this a foolish decision, and said that the three astronauts might never be able to return to earth. *Apollo 8* reached the moon safely, and was able to circle it ten times on Christmas Eve, transmitting television pictures of the moon's surface back to earth. Those television images were broadcast live to homes all over the world. Seated in their living rooms, people on earth were able to gaze at the lunar landscape 239,000 miles away, seemingly near enough to reach out and touch. At the same time we could see earth, gleaming blue in the black void of space. *Apollo 8*'s commander Frank Borman spoke to the viewing audience back on earth:

> This is *Apollo 8* coming to you live from the moon. We've had to
> switch the TV camera now. We showed you first the view of earth as

we've been watching it for the past 16 hours. Now we're switching so we can show you the moon that we've been flying over at 60 miles altitude for the last 16 hours.

The moon is a different thing to each one of us. I think that each one carries his own impressions of what he's seen today. I know my own impression is that it's a vast, lonely, forbidding type expanse of nothing. It looks rather like clouds and clouds of pumice stone. And it certainly does not appear to be a very inviting place to live or work.

Jim [Lovell], what have you thought most about?

Well, Frank, my thoughts were really somewhere in the vast loneliness up here on the moon. It's awe-inspiring and it makes you realize just what you have back there on earth. The earth from here is a grand oasis in the great vastness of space.

The moon was a "dead world," while the earth was a unique "living planet," floating in the black void of space. The photographs of earth, that small blue-green sphere, left a deep impression in the minds of the public, evoking such concepts as "only one Earth" and "Spaceship Earth." The notion of "Spaceship Earth" was originally conceived by the American economist Kenneth E. Boulding, who described it to the Resources for the Future forum on March 8, 1966. Using a clever analogy, he distinguished two basic economic systems. The first he called the "cowboy economy," evoking the image of a cowboy who roamed the vast plains of the American West, with no concern for exhausting what seemed then to be unlimited natural resources. The cowboy connotes reckless, romantic, even violent behavior.

By contrast, Boulding dubbed the other economic model the "spaceman economy." Everything on board a spaceship is limited: water, air, food; all will eventually be used up unless special measures are taken. The carbon dioxide gas and body wastes produced by the occupants of the spaceship will likewise eventually contaminate and pollute the interior of the craft if nothing is done. Boulding was well aware that such an era was far in the future. But many people, as they gazed at *Apollo 8*'s photographs of our small blue-green globe, felt as if the "spaceman economy" might arrive tomorrow.

The following year, on July 20, 1969, the lunar module *Apollo 11* carried spacecraft commander Neil Armstrong and crew member Edwin "Buzz" Aldrin to a safe landing in the Sea of Tranquility on the surface of the moon. The landing area was a desolate, dust-covered desert. Seeing it via television, viewers on earth felt even more certain that only their planet could ever serve as a habitation for living beings.

The success of *Apollo 11* displayed to the whole world the wonders of American science and technology. The United States had at last overtaken the Soviet Union in the race for space. Most of the citizens of America and the rest of the free world welcomed and hailed the lunar landing as the crowning achievement of a tremendous project that had few if any parallels in history. However, there were some who raised dissenting voices against this historic feat. On the day before *Apollo 11* was to be launched from the Kennedy Space Center, about one hundred black demonstrators appeared in front of the Visitors Center. They had gathered from eight southern states, and all carried signs bearing slogans like: "Mr. Nixon: When will you launch some food?" . . . "Countdown: 5, 4, 3, 2, 1—cease fire!" . . . "How can we think about space when our stomachs are empty?"

On July 20, the day Armstrong and Aldrin reached the moon, about 50 blacks staged a demonstration in front of the Manned Spacecraft Center in Houston. They, too, carried signs: "Moon rocks or daily bread?" . . . "Don't forget the hungry!" . . . "Food for the astronauts of the future!"

Leaflets handed to reporters and spectators by the demonstrators read: "Now that the dream of landing on the moon has come true, we want the American dream — a decent life for every American — to come true also." The views expressed by these demonstrators did not contain overt attacks on, or hostility toward, science and technology. But an element of antagonism was certainly present. Among intellectuals and critics, likewise, there were some who expressed open hostility. For example, Lewis Mumford, the well-known author of *The City in History* and *The Myth of the Machine*, put it in these terms:

> The most conspicuous scientific and technical achievements of our age — nuclear bombs, rockets, computers — are all direct products of war, and are still being promoted, under the guise of "Research and Development," for military and political ends that would shrivel under rational examination and candid moral appraisal. The moon landing program is no exception: it is a symbolic act of war, and the slogan the astronauts will carry, proclaiming that it is for the benefit of mankind, is on the same level as the Air Force's monstrous hypocrisy — "Our Profession is Peace."

According to Mumford, the purpose of the moon landing program was to develop weapons for the destruction of mankind, and to preserve what he calls the military-industry-science complex. "It is no accident," he pointed out, "that the climactic moon landing coincides with cutbacks in education, the bankruptcy of hospital services, the closing of libraries and museums, and the mounting defilement of the urban and natural environment." He continued:

> If a successful moon landing leads to a further expansion of space exploration with a further drain on more important human enterprises and a further neglect of the conditions essential for human survival and development, we may look forward to a corresponding increase in social demoralization and psychological regression. Only a return to full waking consciousness, with an overwhelming transfer of interest from our dehumanized technology to the human person, will suffice to bring our moonstruck nation back to earth. Meanwhile, thanks to the very triumphs of technology, the human race hovers on the edge of catastrophe.

These views of Lewis Mumford appeared in the *New York Times*, right alongside comments by others in praise of the lunar landing. Japan's *Asahi Shimbun* also carried comments with an anti-scientific tone. For example, Taruho Inagaki wrote:

> I, for one, hope that this latest outbreak of technological insanity will be brought under control by mankind's innate imagination and powers of thought. Wasn't the original Apollo the patron deity of harmony and order?

Similarly, another author, Mrs. Aiko Sato, noted:

> I am a conservative sort of person, and when civilization goes this far, I
> feel as if I am being flung about, and I'm beginning to be afraid.
> Wouldn't it be better just to call a halt, for now, to any further de-
> velopments of this kind?

Apollo 11 returned safely to earth on July 24 — about one week after the unmanned
space probe *Mariner 6* passed close to Mars. The close-up photographs it transmitted
back to earth showed large numbers of craters, similar to those on the moon. There
were predictions that Mars, like the moon, would turn out to be a dead world. Soviet
probes of Venus had already clearly indicated the surface of that planet to be an
inferno of extreme heat. Now that even Mars was proving to be a "planet of death," it
became clearer than ever before that our own earth might be the only planet on which
human beings could ever survive. People began to turn their eyes back toward their
earth. In doing so, many members of the public began to believe, as many intellectuals and critics were charging, that science and technology might be ruining the
earth. There spread throughout the free world a fear that science and technology were
leading mankind and the planet toward destruction.

Around that time, the word "ecology" began to appear frequently in the press and
other media. Ecology, the study of the relationship between living things and their
environment, had originated in the mid-nineteenth century; but developments in
ecological research had been overshadowed after the second World War by advances
in newer biological specialties like analytical biochemistry, molecular biology, and
biophysics. The publication of Rachel Carson's *Silent Spring* breathed new life into
ecology, and the defoliation campaign in Vietnam focused worldwide attention on the
idea of ecological problems.

For ecology, each instance of environmental damage or pollution must be treated
not as an isolated problem, but as one that involves all of man's environment and the
natural world as a whole; the skillful use of the natural enemies of harmful insects,
rather than pesticides, often proves effective in controlling or eliminating them; and
once a particular ecosystem has been destroyed, many hundreds or even thousands of
years may be needed to restore it.

One other major scientific and technological advance played a role in science
thinking around the same time: the significant increase in the degree of accuracy and
precision achievable by chemical analysis. The analytical technique called chromatography had been developed in 1952, and about ten years later, the analytical
procedure known as gas chromatography came into use. With this technique one is
capable of detecting and measuring extremely tiny traces of chemical substances, as
little as one part per million or less. Minute amounts of impurities and contaminants
— undetectable prior to the development of these new techniques — were suddenly
discovered. Terms like "parts per million" (ppm) and "parts per billion" (ppb)
appeared more and more frequently in the press. Now a newspaper article might note
that "analysis has revealed that cow's milk contains 2.68 ppm of the insecticide
BHC." With these advances in analytical techniques, the general public learned for
the first time that cow's milk and human milk were contaminated by pesticides. But
instead of being grateful to science for these new procedures, people simply attacked
science and technology for making the pesticides. Thus, scientific and technological

progress ironically succeeded in making a major contribution to the worsening of the public's appreciation of science.

President Nixon and Earth Day

The person who, at one stroke, sharply aggravated this anti-science sentiment was President Richard M. Nixon. On January 1, 1970, he signed the National Environmental Policy Act, stating that "it is particularly fitting that my first official action in this new decade is to approve the National Environmental Policy Act." In his informal remarks at the signing ceremony, he remarked that the task "for the next ten years for this country must be to restore the cleanliness of the air, the water." President Nixon singled out the environmental issue as one of the major policy concerns for the decade of the 1970's:

> If you look ahead ten years, you project population growth, car growth, and that means, of course, smog growth, water pollution, and the rest. An area like this will be unfit for living; New York will be, Philadelphia, and of course, 75 percent of the people will be living in areas like this.

He then emphasized that the problem was not America's alone, but one confronting the entire world:

> Then when you look at it, too, I have noted in all my conversations with the heads of government of the major industrial nations — for example, Sato in Japan, Wilson in England, the German leaders, the new Chancellor, Brandt, the French leaders, the Italians, and all the rest — all of them have similar problems.

President Nixon explicitly declared his intention of tackling environmental problems from a worldwide standpoint, and even appealed to the United Nations. In 1972, the United Nations decided to hold an international conference on environmental issues in Stockholm. The movement to clean up the environment and protect nature thus became a worldwide political issue, and policy decisions in this area contributed greatly to making the human environment healthier, safer, and more attractive.

However, Nixon's motives were not totally idealistic. Before his inauguration as President, he had already created a number of task forces to study possible policies to be adopted by the new administration. One of these, the Task Force on Resources and Environment, was chaired by Russell E. Train, President of the Conservation Foundation. It presented its report on environmental problems to the President-elect on December 5, 1968. The report began by describing the deterioration of the environment, and then pointed out that during the 1960's:

> Again and again, in state and local referendums across the nation, voters have given their approval — often by lopsided margins — to bond issues for open space acquisition, outdoor recreation programs, pollution abatement. . . .Determined and effective citizen opposition to freeways, dams and loss of natural areas is commonplace.

On the day of the 1968 presidential election a Michigan proposal to spend 335 million dollars on water pollution control was approved by 70 percent of the voters. In the state of Washington, 72 percent of those voting approved a plan to spend $40 million on recreational land acquisition and development. Such figures made it abundantly clear that environmental issues were big "vote-getters" at election time. The report went on:

> Internationally, these problems constitute an extraordinary opportunity for United States leadership and new initiatives. Environmental quality is a unifying goal that cuts across economic and racial lines, across political and social boundaries. It is a goal that provides a new perspective to many national problems and can give a new direction to public policy. Its values and support come not from the divisions that plague our society but from the common aspirations of all for a life of dignity, health and fulfillment.

In 1968, Vietnam was a major issue with the American public. There were other equally important domestic concerns like racial discrimination and economic inequality. The report seemed to suggest that policies emphasizing environmental issues might be able to defuse or displace these other, more divisive, domestic and foreign problems. However, the hard core opponents of the Vietnam war, as well as the blacks who demanded an end to racial discrimination, were quick to see through President Nixon's intentions. They began objecting to his attempts to distract the public's attention with environmental concerns. Black leaders denounced the environmental issue as an expensive hobby for wealthy whites.

The President of the United States, nonetheless, is an extremely influential figure, and his statements carry great weight in every other country of the free world. To take just the area of science and technology, for example, there was President Eisenhower's announcement of the "Atoms For Peace" program, in late 1953. The very next year, an "atomic energy boom" sprang up in Japan. Bureaucrats, scientists, politicians, businessmen — everyone in Japan seemed to be chasing after atomic energy, as if afraid to be the last one aboard the train. Again, in 1961, when President Kennedy announced the Apollo space program, Japan also experienced a "space boom." Later, President Johnson (1965) proposed to apply space technology to the world's oceans, and Japan dutifully underwent an "ocean boom," and a whole series of ocean development companies were organized. Thus, President Nixon's environmental policy had an immeasurably important effect not only on America, but on other countries as well.

In America, only four months (April 22) after President Nixon's inauguration, "Earth Day" was observed for the first time. As the date happened to be also Lenin's birthday, there were some who suspected a Communist conspiracy, but the undertaking proved to be a great success. The *New York Times* reported the event under the big headline, "Millions Join Earth Day Observances Across the Nation." Some of those in the crowds wore gas masks and carried signs reading "Don't Breathe!" Others held up signs showing a picture of a nursing mother's breasts, captioned "Caution, Keep Out! DDT 0.3 PPM!"

The major success scored by the Earth Day gatherings was in large part due to the fact that the "global awareness" of the public had already been heightened by the photographs taken by the Apollo astronauts, as I have already said. But it is also certain that President Nixon's environmental statements and policies had the effect of swelling the number of participants. Few blacks, however, took part in the Earth Day observances. The following day, the *New York Times* observed in an editorial:

> Is the sudden concern for the environment merely another "nice good, middle-class issue," as one organizer put it, conveniently timed to divert the nation's attention from such pressing problems as the spreading war in Indochina and intractable social injustice at home?
> We think not. . . .
> If Earth Day has diverted the energy of Americans from other causes it is because many have finally perceived that the problems of the environment also have an urgent claim on national attention. . . .
> It is also self-evident that pollution does not discriminate. The environment encompasses all Americans, for better or for worse — white and black, rich and poor, right and left. Unless all can live and work together for a better environment, all may suffocate together.

The mass media adopted a cooperative attitude toward President Nixon's environmental program, and the problems of pollution and the environment began to receive major coverage in the press, radio, and television. The *New York Times* and other newspapers hastily created positions for environmental reporters and editors.

Nevertheless, just as almost every drug has side effects of some kind, President Nixon's environmental policy also had its own serious side effect: it greatly aggravated popular trust in science. On a number of occasions, the President indicted science and technology as the chief culprits behind the deterioration of the environment. On August 10, 1970, for example, when he submitted the first annual report of the Council on Environmental Quality to Congress, President Nixon commented that "rapid population increases here and abroad, urbanization, the technology explosion and the patterns of economic growth have all contributed to our environmental crisis."

Most Japanese believed that whatever the American people did was excellent in every respect. The Japanese people, accordingly, diligently imitated everything done by the Americans. Those Japanese who knew the English language translated American books and articles, visited America, and assiduously introduced America's superb culture, science, and technology into Japan. Newspaper writers were among those eager to "import" America's advanced culture into Japan. I myself, as I have already remarked, came to the United States several times as a special correspondent for the *Asahi Shimbun*, sending back articles about America's progress in nuclear power and space technology.

One actual and rather typical incident will, I think, suggest how Japan looked up to American knowhow. A young researcher at the main laboratory of a large Japanese manufacturing firm came up with a new idea. He wrote a request to the director of the laboratory asking for permission to start research to see if the idea was commercially feasible. The director asked, "That idea was developed in America, wasn't it?" "No," replied the younger man, "I came up with it on my own. I don't believe that they've developed a product like it in America yet." The young researcher eagerly explained how original his idea was, but the director did not seem pleased. "That's no good. If it hasn't been done in America, how could it work when *you* do it?"

It was in an atmosphere like this that Japan's "atomic energy boom" occurred. Accordingly, when President Nixon chose environmental issues as a policy focus for the decade of the 1970's, it was only natural that a similar "environmental boom" should occur in Japan. Simultaneously, the notion voiced by President Nixon and others — that science and technology had brought about a crisis in the environment — was imported directly into Japan. In this manner, the dangerous pathogens of the anti-science disease were transmitted to Japan.

Japan offered these germs a fertile environment in which to multiply, because America's bout with the anti-science disease had already been experienced in Japan in the spring of 1969. Ralph Nader's *Unsafe at Any Speed* produced almost immediate repercussions in the United States, but attracted little attention in Japan. This was primarily because Japan had not yet become a major manufacturer and exporter of automobiles. Japan produced only 1.88 million cars in 1965, about one-sixth of America's annual total, and there were very few high-speed highways in Japan. But by 1969, the number of Japanese automobiles produced annually had climbed to 4.67 million, 2.5 times as many as in 1965. Conditions in Japan were finally ready for a Nader-style indictment of defective automobiles, which began with an article in the *New York Times*. It appeared on Page 34 of the May 12, 1969 issue of the *Times* under the smallish headline, "Publicity Avoided on Some Recalls." The story read:

> Safety recall campaigns are conducted without public announcement by five of the 10 leading importers of foreign cars, a check by the *New York Times* has disclosed.

Following Nader's revelations about unsafe automobiles, new Federal legislation had been passed in the United States in 1966. Under this law, whenever a safety-related defect was thought to exist, the manufacturer was required to notify all owners of the affected model by certified mail, recall the vehicles, and make the needed repairs or modifications. The law did not oblige the manufacturer to make a public announcement in the event of a recall, but American auto companies nevertheless invariably did make such an announcement. Foreign manufacturers, however, including Japan's Toyota and Nissan (Datsun), did not give public notice of the fact of a recall. The *New York Times* article continued:

> Recent unannounced and unpublicized recall campaigns included one started last month by the Nissan Motor Corporation, involving 39,425 cars. All are 1969-model Datsuns. The notice filed with the National Highway Safety Bureau listed possible leakage in the fuel pump and carburetor, either of which defect could cause fire.
>   Another recent recall, likewise unannounced and unpublicized, was instituted last January by Toyota Motor Sales, Inc., for 18,395 Toyota Corona cars of the 1969 model. A possibly faulty seal in the brake fluid reservoir, which could cause leakage and loss of braking ability, was listed.

Japan's Kyodo wire service distributed a Japanese translation of this article, which created an immediate outcry in Japan. The Japanese newspaper editorial commentary ran along these lines: Automobiles identical to the recalled models are operating on Japan's streets and highways, but the Japanese manufacturers, far from notifying the owners, have even concealed the defects from the companies' own employees — they are ignoring the safety of drivers and passengers in Japan, and their sales efforts are seemingly being allowed to run wild.

The automobile had suddenly become a "suicide machine on wheels," a "rolling coffin" in Japan. Even the transportation committee of the Lower House of Japan's parliament pursued the issue vigorously. The tone of the denunciations grew more violent day by day, and escalated to assertions in newspaper articles that "one Japanese car in ten is defective with 1.3 million defective cars now on the road," or that "drivers are treated simply as guinea pigs."

These accusations about defective cars proved extremely effective in improving automotive safety, as manufacturers began to pay much greater attention to safety factors. However, just as Ralph Nader had increased America's susceptibility to the anti-science disease, so the indictments of faulty automobiles in Japan aggravated the tendency to distrust technology.

About four months later (October 1969), the ban on the artificial sweetener called cyclamate created a popular sensation. Again, the point of origin of the disturbance was the United States. On October 18, Robert H. Finch, Secretary of Health, Education, and Welfare, announced the ban on the grounds that experiments performed on laboratory mice had proven that cyclamate was a carcinogen.

Because cyclamate has no bitter after-taste (unlike saccharine) and is thirty times sweeter than sugar, it was used at the time for about 80 percent of all artificial sweetening. It was a familiar substance, widely used in soft drinks, canned fruit, ice cream, jams and jellies, condiments, and so on. The word "carcinogen," however, has the power to send shudders through the public, to cause panic and hysteria. Moreover, this incident occurred at a time when, as I have explained, most Japanese believed that any American decisions on key issues should be taken as a model for Japan. Japanese newspaper readers saw headlines like "Total U.S. Ban on Man-made Sweetener Cyclamate; Possible Cause of Cancer," and reacted in a virtually hysterical manner. Japan's Minister of Health and Welfare issued a statement: "We intend to obtain the actual laboratory test data from the United States, study it carefully, and then make a decision on whether or not to ban the use of cyclamate in Japan."

Public opinion, however, would not stand for such a leisurely approach to the problem. Newspapers ran headlines about cyclamate: "No Restrictions in Japan: Used in Most Inexpensive Foods." Reading these, the public began to have the feeling that this was an emergency situation, one that demanded immediate attention. The Minister of Health and Welfare, intimidated by the sudden flare-up of public opinion, did not even wait for the American laboratory data to arrive before asking the food industry to exercise "voluntary restraint" in their use of cyclamate. Fruit juices and canned foods containing cyclamate remained on store shelves, and mountains of returned and rejected goods were buried by bulldozers in huge pits. It is now known that the experiments that declared cyclamate a carcinogen, and thus started this panic, were in error; but at the time, the public was swept along by a strong, emotional distrust of any substance that, like cyclamate, was created by scientific means. This emotion was easily carried over into a suspicion of science and technology more generally.

This uproar over cyclamates, which began in America, played a major role in creating a basis for the anti-science sentiment in Japan, a basis upon which President Nixon's policies built. The data at the end of Chapter III suggested that the outbreak of the anti-science disease in Japan occurred around 1970. Japanese newspapers at this time carried reports of environmental pollution on virtually a daily basis, as well as frequent special features. In these writings, as might be expected, science and technology came under attack. The *Asahi Shimbun*, for example, ran a special feature page called "Let's Defend the Environment," containing an essay, by social psychologist Hitoshi Aiba, entitled "Civilization Made Sick by the Gods and Demons Science Harbors." Aiba, in summarizing American anti-science views, stirred up Japanese sentiments:

> Just as both God and the Devil are said to inhabit Creation, so we have begun to discover that science also has its satanic aspects. . . .Automobiles emit poisonous exhaust gases. Factories spew forth pollutants. Even the most commonplace and familiar things in our daily lives are being used by science and technology to warp and twist mankind. Americans seem to have taken notice of this satanic side of science and technology, and are deeply aware of it. America is said to be ill — but from the very depths of America in her illness can be heard voices, asking to be given back the health that was twisted and distorted by scientific civilization.

The Limits to Growth

The American Newspaper Publishers' Association held its 1970 annual convention in April at the Waldorf-Astoria Hotel in New York City. A view expressed with virtual unanimity by the assembled editors and publishers was that "Vietnam" and "racial discrimination" had lost their appeal as issues by 1969, and that people were now deeply concerned with environmental pollution. Anti-war marches had diminished in number and resistance to discrimination had weakened in strength. In their place, pollution and environmental destruction had become major topics. The political aim of President Nixon and his advisers — to divert the American people's attention from Vietnam and racial problems — had been brilliantly accomplished. As issues, war and discrimination had been swept away by an "environmental boom." An important manifestation of that boom was the high praise accorded to books that further aggravated the anti-science disease.

One of the most important of these books was *The Population Bomb*, by Dr. Paul R. Ehrlich. It was published in May 1968 and went through three printings during that year alone. Seven more printings followed in 1969 and twelve more in 1970, for a total of 22 printings in less than three years. Like *Silent Spring*, Ehrlich's book was a chronicle of horrors. The principal themes of the book were that the population of the earth is steadily increasing, the environment is being destroyed, and the planet is headed for catastrophe. The book's prologue began with the words:

The battle to feed all of humanity is over. In the 1970's and 1980's hundreds of millions of people will starve to death in spite of any crash programs embarked upon now.

Ehrlich insisted that the earth's rate of population growth must be reduced to zero. Responding to this theme, an organization known as Zero Population Growth (ZPG) was formed; membership had passed the 30,000 mark by February 1, 1971. Ehrlich's book included elements that tended to encourage the anti-science disease:

It is probably in vain that so many look to science and technology to solve our present ecological crises. Much more basic changes are needed, perhaps of the type exemplified by the much-despised "hippie" movement — a movement that adopts most of its religious ideas from the non-Christian East. It is a movement wrapped up in Zen Buddhism, love, and a disdain for material wealth.

In March 1972, an explosive book called *The Limits to Growth* was published. It was a report on "the predicament of mankind," compiled by a Massachusetts Institute of Technology project team, at the request of the Club of Rome. The Club of Rome is a non-governmental organization founded in March 1970, consisting of distinguished scientists, economists, educators, and business executives from a number of different countries. Created in response to a proposal by Dr. Aurelio Peccei, the vice president of the Italian business machine firm Olivetti, the group's first meeting was held in Rome, from which it took its name. In June 1970, the Club of Rome began its "Project on the Predicament of Mankind," and in July of that year commissioned the study by the team at M.I.T. The project began after President Nixon announced his environmental stand, and for that reason it might be appropriate to see the study's conclusions as a "child of the times." The book warned:

If the present growth trends in world population, industrialization, pollution, food production, and resource depletion continue unchanged, the limits to growth on this planet will be reached sometime within the next one hundred years. The most probable result will be a rather sudden and uncontrollable decline in both population and industrial capacity.

The work read like a compilation of numerous individual views at the time. One novel feature was that all available data had been entered into a computer in order to predict the future, and this new approach induced many people to buy and read the book. According to the foreword of the Japanese language edition, the initial American printing of 20,000 copies was sold out in one week, and the Dutch edition also sold out all 10,000 copies in a week. The book was an extremely pessimistic one:

The hopes of the technological optimists center on the ability of technology to remove or extend the limits to growth of population and capital. We have shown that in the world model the application of technology to apparent problems of resource depletion or pollution or food shortage has no impact on the essential problem, which is exponential growth in a finite and complex system. Our attempts to use even the most optimistic estimates of the benefits of technology in the model did not prevent the ultimate decline of population and industry, and in fact did not in any case postpone the collapse beyond the year 2100.

The Limits to Growth insisted upon the impotence of technology:

> Technology can relieve the symptoms of a problem without affecting the underlying causes. Faith in technology as the ultimate solution to all problems can thus divert our attention from the most fundamental problem — the problem of growth in a finite system — and prevent us from taking effective action to solve it.

At the end of the book was a "commentary" by the Club of Rome:

> It was suggested that insufficient weight had been given to the possibilities of scientific and technological advances in solving certain problems, such as the development of foolproof contraceptive methods, the production of protein from fossil fuels, the generation or harnessing of virtually limitless energy (including pollution-free solar energy), and its subsequent use for synthesizing food from air and water and for extracting minerals from rocks. It was agreed, however, that such developments would probably come too late to avert demographic or environmental disaster. In any case they probably would only delay rather than avoid crises, for the problematique consists of issues that require more than technical solutions.

The photographs taken by the Apollo spacecrafts had shown us, in a visual and emotional sense, the boundaries of our planet; but *The Limits to Growth* made us aware of these limitations in a logical and rational manner. Yet, like *Silent Spring* and *The Population Bomb*, the Club of Rome's book was a chronicle of disaster. Readers were frightened and intimidated by this computerized oracle announcing that a major catastrophe would occur within the next hundred years. According to the book's forecasts, which may seem exaggerated today, all the earth's known reserves of gold would be exhausted in nine years, those of silver and mercury in 13 years, and those of lead in 15 years. Counting from the book's year of publication (1972), twelve years have already elapsed, and the thirteenth and fifteenth "years of exhaustion" are just around the corner. Yet we have not yet exhausted resources for any of those metals. *The Limits to Growth* may have been a "chronicle of overstatement," well-attuned to the fashions of its day. The book not only strengthened certain ideas in vogue at the time, but also made the public anti-science sentiment that much worse. Words like "limits" and "boundaries" became popular catchwords. Even among scientists and engineers, there were some who thought and talked as if science and technology had already reached the "limits" of progress. For example, Dr. Bentley Glass, a biologist serving as president of the American Association for the Advancement of Science, made the following remarks in December 1970, at the end of his term in office:

> The great conceptions, the fundamental mechanisms, and the basic laws are now known. For all time to come these have been discovered, here

and now, in our own lifetime. We are like the explorers of a great continent who have penetrated to its margins in most points of the compass and have mapped the major mountain chains and rivers. There are still innumerable details to fill in, but the endless horizons no longer exist.

As a biologist, Glass could make those comments when he recalled the explosive development of a field like molecular biology in the 1960's. The Japanese biologist Atsuhiro Shibatani wrote in his book *The Theory of Anti-Science*:

Perhaps never again will science experience anything like the golden age of the early 1960's, when (for example) molecular biology first came into full flower.

Predictions to the effect that "progress has come to an end" have consistently been proven wrong. The American physicist Albert Michelson, for example, made the following declaration in 1892:

It seems probable that the grand underlying principles of physical science have been firmly established and that further advances are to be sought chiefly in the rigorous application of these principles to all phenomena. An eminent physicist has remarked that the future truths of physical science are to be looked for in the sixth place of decimals.

The famous British physicist Lord Kelvin is said to have made similar statements at that same time. But within a few years of these pronouncements, the field of physics witnessed such major developments as quantum mechanics, relativity theory, X-rays, radium, and nuclear fission. From the latter part of the 1960's and well into the 1970's, a considerable number of scientists and engineers tended toward the pessimistic view that science and technology had come to the end of the road, that there would be little future progress. But today, science and technology are moving forward by leaps and bounds. Major progress is being achieved in a host of scientific and technical areas: computer miniaturization; gene manipulation to create new medicines and crops with higher yields; communication using laser beams and fiber optics instead of radio waves or electrical currents; and the unlocking of the mysteries of space by means of astronomical satellites.

One wonders what biologists like Glass and Shibatani, who once declared that the golden age of science was over, would now say about the field of genetic research, which is currently advancing at a tremendous pace. Looking back from the present, it is easy for us to see how mistaken they were — but at the time, many people shared their views. The American physicist Dr. Harvey Brooks, who once served as a member of the President's Science Advisory Committee, has commented on Glass's views:

Glass's views are frequently echoed among other scientists. Although expressed mostly by the older generation of scientists, one also finds young people turning away from science for essentially these reasons, and this view of the finite compass of discoverable knowledge is likely to erode interest in science.

V
DELAY, STUMBLE, FALL

Groups that catch the anti-science disease often try to impede developments in science and technology by organizing and encouraging anti-technology crusades. For example, America's supersonic transport (SST) which met an untimely end, was the victim of just such a campaign. Voices from the American aircraft industry advocating development of an SST began to be heard in the late 1950's. The Federal Aviation Administration (FAA) set up an SST Study Group in December 1959, and in March 1960 the SST Planning Group was created. In May of that year, a subcommittee of the House of Representatives held hearings on the SST. At that point, industry experts were extremely optimistic about the prospects for the transport. The construction of an SST was quite feasible from an engineering standpoint, and the commercial introduction of the new aircraft would have been possible sometime between 1965 and 1970. Most experts asserted, moreover, that America should build a passenger transport capable of flying three times the speed of sound (Mach 3).

Around that same time, Britain and France were holding talks to consider joint development of a passenger aircraft that would fly at Mach 2. An aircraft designed to fly Mach 2, it was thought, could be built with an aluminum alloy, like conventional subsonic jet aircraft. To fly at Mach 3, however, it would be necessary not only to "break the sound barrier," but to penetrate the so-called heat or thermal barrier as well. Because of the friction generated by the air molecules as the aircraft forced its way through the atmosphere at tremendous speeds, the nose of the plane and the leading edges of its wings would be heated to temperatures around 400 degrees Fahrenheit. The nose and the wing edges, therefore, had to be made of titanium alloy and stainless steel. This in itself was still difficult from a technical standpoint. America, however, did not want to be beaten by the British and the French. The Soviet Union, moreover, was also expected to attempt to develop an SST, and the United States certainly had no desire to lose out to the Russians. For that reason, many American experts wanted to build an SST that could fly at Mach 3. Most of the commercial jet transports in the world's skies today reach speeds of between Mach 0.7 and Mach 0.9. Since they fly close to the speed of sound, but do not exceed it, they are referred to as "transsonic" transports. Such an aircraft can fly from Los Angeles to Tokyo in about 14 hours. A Mach 3 SST, however, could make the same journey in only five hours. A trip across the Pacific Ocean and back in only one day would no longer be an impossible dream.

While the American experts were debating whether to try for a Mach 2 or a Mach 3 aircraft, the British and French reached an agreement, and in November 1962 the two countries announced a program for the joint development of an SST, to be named the *Concorde*. At that time, FAA Administrator Najeeb Halaby sent President Kennedy a memorandum stating that the United States must also develop an SST. He declared that unless America did so, it would lose its position as the world's leader in aviation and aircraft manufacturing. Aircraft export sales would decline causing trade deficits. Moreover, he wrote, the President of the United States might someday be forced to travel abroad in a foreign-built *Concorde* aircraft. In January 1963, President Kennedy asked the FAA to prepare a report on the SST. The report was completed in May. On the basis of that report President Kennedy stated, in a graduation address at the U.S. Air Force Academy on June 5, that he intended to emphasize development of an SST as a national program:

> It is my judgment that this Government should immediately commence a new program in partnership with private industry to develop at the earliest practical date the proto-type of a commercially successful super-sonic transport superior to that being built in any other country of the world. . . .If we can build the best operational plane of this type — and I believe we can — then the Congress and the country should be prepared to invest the funds and effort necessary to maintain this Nation's lead in long-range aircraft, a lead we have held since the end of the Second World War, a lead we should make every responsible effort to maintain.

On June 4, 1963, one day before President Kennedy's speech, Pan American Airways announced that it intended to purchase six *Concorde* SST's. The President had been notified of this announcement in advance, and it presumably hastened his decision to press forward with the American SST project. He was concerned that unless the United States went ahead with the SST, most of the world's airlines would end up deciding to buy the *Concorde*. Thus, the development of an American SST became an official national objective. Research and design for the aircraft, however, did not proceed nearly as smoothly as planned. For example, computations showed that the fuselage design was far heavier than expected, making the aircraft too heavy even to leave the ground. The plane had to be completely redesigned. Development costs for the SST, under the guidelines initially set forth, were to have been shared by the Federal government (75 percent) and the manufacturers (25 percent). The latter, however, claimed they could not afford to pay 25 percent of the costs, so the government ended up paying 90 percent and the manufacturers 10 percent.

While the program dragged along, America was questioning its faith in science. I noted in Chapter III that the symptoms of America's illness became evident around 1965, and the first voices raised in protest against the SST began in 1966. In July of that year, *Harper's* magazine published an article by Dr. John E. Gibson, titled "The Case Against the SST." The July 28, 1966 issue of the *Wall Street Journal* also ran an editorial opposing development of the SST. In August 1966, Senator William Proxmire (Democrat from Wisconsin) proposed that the $280 million set aside for the SST in the fiscal 1967 budget be cut to only $80 million. His proposal was rejected by a Senate vote of 55 to 31, but this was the first chapter in the story of Senator Proxmire's opposition to the SST.

Around this time Dr. William A. Shurcliff, a physicist at the Cambridge Electron Accelerator of Harvard University, took an interest in the issue. He had read in the February 1965 issue of the *Bulletin of the Atomic Scientists* an article by Dr. Bo Lundberg, then Director General of the Aeronautical Research Institute of Sweden. The article carried a strong warning about the "sonic booms," or shock waves, that supersonic aircraft had been creating since 1960. Shurcliff later said that from the moment he read that article, he began to be concerned about sonic booms.

In the latter part of 1966, the *New York Times* began to carry letters to the editor expressing opposition to the SST. One of these letters was written by Dr. John T. Edsall, a prominent biologist at Harvard University, who had become known for his vigorous opposition to the Vietnam defoliation campaign. After Shurcliff read the letter in the *New York Times*, he went to see Edsall, and they exchanged views on the SST issue. The two scientists founded the Citizen's League Against the Sonic Boom (CLASB) on March 9, 1967, and ran large half-page advertisements about the dangers of the SST and its sonic boom, in two issues of the *New York Times*. As a result of this publicity, several hundred persons applied for membership in CLASB, and around $10,000 was contributed to the cause. Shurcliff used these funds to run similar advertisements in the *Christian Science Monitor*, *Wall Street Journal*, and *Washington Post*. At the same time, Shurcliff kept writing letters to government officials, members of Congress, newspapers, and magazines.

The headquarters of CLASB was a small room in Shurcliff's home, and the only activists in the organization were Shurcliff himself, his daughter, and his son. What enabled this tiny group ultimately to bring down the SST was Shurcliff's own strategy, which concentrated on the skillful use of the mass media. He kept sending material to about 200 newspapers all over the United States. Many of these papers used his material as the basis for news stories, and it is said that in 1967 and 1968, an average of five articles per day appeared in newspapers across the country.

For activists in opposition movements, the mass media provide the most effective weapon. Journalists love anything out of the ordinary — as the old saying has it, "When dog bites man, that's nothing, but when man bites dog, that's news!" When things are going along uneventfully, the newspapers and television never report that fact — but when it comes to a story involving a controversy or some sort of "danger," journalists get a gleam in their eyes and rush off to cover the story.

The views of Shurcliff and his supporters received frequent media coverage in 1969. In September of that year, however, President Nixon (who had taken office in January 1969) announced a plan to give new impetus to the SST program. His reasons were largely identical to those expressed earlier by President Kennedy. "I want the United States to continue to lead the world in air transport," Nixon declared. The fiscal 1970 budget for the SST was uneventfully approved by Congress on December 26, 1969. However, a heated debate broke out in Congress about the fiscal 1971 appropriation, with the anti-SST legislators being spearheaded by Senator Proxmire. A number of witnesses were summoned to testify before Congressional committees. Opponents of the program took various tacks: the noise from one SST taking off would be equal to that of 50 "jumbo jets"; people living 15 miles away from an airport used by SSTs would suffer from noise as loud as if they lived only one mile from an airport handling conventional jets; and whenever an SST flew overhead at supersonic speed, window glass would be shattered by the continuous sonic wave the plane generated.

The testimony with the most powerful impact, however, came from the chairman of the Council on Environmental Quality, Russell E. Train. Train had served as chairman of President Nixon's pre-inaugural Task Force on Resources and Environment, and had recommended to the President a policy of emphasizing environmental issues. Train testified before Senator Proxmire's Joint Economic Committee in May 1970, and voiced the theory that the SST would cause an increase in skin cancer. The logic of this theory ran as follows. The exhaust gases from the SST contain steam, which is water vapor. If 500 SSTs are eventually put into operation, the result will be an increase of between 50 and 100 percent in the amount of water vapor in the upper layers of the earth's atmosphere. One of these layers is composed of a gas called ozone, which will be destroyed by the added water vapor. The ozone layer, however, plays a vital role in absorbing severe ultraviolet radiation emitted by the sun, and if there is less ozone to absorb these rays, more intense ultraviolet radiation will strike the earth's surface. Too much ultraviolet light increases a human being's likelihood of developing skin cancer. Thus, the conclusion to be drawn was that the SST would increase skin cancer.

Train, however, had absolutely no scientific evidence to support his conclusion. His testimony was nothing more than a hypothesis — in fact, it belonged more to the realm of speculation. The nature of the earth's ozone layer was not fully understood, and no one really knew whether or not the water vapor emitted by an SST would actually destroy the ozone layer. Furthermore, the possibility of an increase in skin cancer was sheer conjecture. American newspapers and magazines, however, failed to point out these flaws in the theory, and instead gave big headlines to the Train hypothesis. "Cancer" is certainly a scare-word, and people were thoroughly frightened by the press accounts.

It was the same with the noise and sonic boom issues. The SST is of course able to fly at less than the speed of sound, so it can reduce its speed below Mach 1 when flying at low altitudes, or taking off or landing at an airport. It has already become widely known that the Anglo-French *Concorde* SST is surprisingly quiet. Fears of noise 50 times louder than a jumbo jet have proven totally unfounded, while there have been no complaints whatever about broken windows caused by the *Concorde*'s sonic boom. In 1970, however, Congress came down with a severe case of anti-science disease. By a vote of 215 to 20, the House of Representatives on March 18, 1971, rejected the budget item for SST development. Subsequently, the Senate also voted against it by a vote of 51 to 46 on March 24, 1971. President Nixon believed that development of the SST ought to continue, but it had run into anti-technological turbulence attributable in large part to the President's own environmental program. One might say that President Nixon brought the SST down on his own head. The "environmental boom" had aggravated America's anti-science disease. The illness was now running wild, beyond the President's control.

The influence of William Shurcliff's anti-SST movement was felt as far away as Japan. Japan Airlines (JAL) had entered into a contract with the Anglo-French company to purchase three *Concorde* SSTs, but later cancelled this contract when the force of the American anti-SST campaign became apparent. If SSTs were not to be allowed to land in the United States, JAL reasoned, then there would be no way to utilize them in a profitable manner. To this day, Britain and France have been unable to sell *Concordes* in Japan.

The campaign by Shurcliff and his supporters, moreover, had a considerable effect upon the construction of the new Tokyo International Airport (Narita). The Japanese government had foreseen that the existing airport at Haneda in Tokyo would become inadequate to handle the annual increases in the volume of air traffic. Accordingly, the government decided in April 1966 to build a new airport in a rural area near the city of Narita, about 35 miles east of central Tokyo. The initial portion of the airport was supposed to commence operations in 1970, so that Japan would be ready to greet the dawn of the SST era. However, beginning in February 1968, a violent campaign of opposition to the new airport began in Japan. This movement, whose members included radical leftist students, succeeded in delaying the opening of the already completed airport for eight years, until May 1978. The motives behind the opposition to the airport ranged from resistance by farmers to the expropriation of their land, to fears that the airport might be put to military use. But one of the major arguments used by the airport's opponents would be the noise and sonic boom of the SST.

Ultimately, however, it was not environmental but economic arrows that shot down the SST. Designed in the fuel-cheap 1960's with only 100 seats and very high fuel consumption, the *Concorde* SST could not operate economically in the era of high oil prices. The Anglo-French company, after producing only sixteen SSTs, of which many were still unsold, shut down its assembly line in September 1979. Only Air France and British Airways currently operate passenger SSTs.

Wait a Minute!

Other technological development programs besides the SST began to encounter delays or cancellations as a result of America's questioning of science. Secretary of Commerce, Maurice H. Stans, apparently unwilling to let this pass unnoticed, delivered a major address before the 25th anniversary meeting of the National Petroleum Council in Washington, D.C., on July 15, 1971. He chose the title "Wait a Minute" for his talk, which began:

> I deeply believe there is a need to review some of the developments that have led business into public and political torment today. They stem, in large part, from a mood of impatience, of frustration, of concern that now pervades much of the country, on many subjects.

Stans then noted that President Nixon had made environmental issues a major policy concern of his administration, and continued:

> But understandable as is the public impatience for solutions, we have the obligation also to see our problems in the whole, not piecemeal. Most of the matters troubling us are interrelated, and we must be aware that if we settle for quick, partial solutions to one set of problems, we can catapult ourselves into other ones far more serious. So we must begin to look a little further down the road. It is high time for the entire nation to weigh needs against demands and say, "Wait a Minute. What are our priorities?" We must weigh requirements

against resources and say, "Wait a Minute. Which can we achieve?"
We must weigh technological capabilities against timetables and options and say, "Wait a Minute. How can we get there from here?"
We must weigh environmental goals against economic reality and
say, "Wait a Minute. How do the benefits of one affect the cost of
the other?"

American industry, he declared, had begun to spend large sums to clean up the air, the water, and the landscape. He listed a number of examples: the chemical industry had spent $600 million for pollution abatement in 1970, and the iron and steel industry had spent more than $1 billion to prevent air and water pollution. Automobile manufacturers were spending $250 million annually, he said, on air pollution research. The electric utilities would spend two-thirds of a billion dollars on pollution control during 1971 alone. America's industries as a whole, Stans stated, would have to spend some $18 billion over the next five years in order to comply with newly-established environmental standards. "Unfortunately," he added, "business has failed to make its achievements and commitments credibly known to the American people."

Stans then cited several actual cases where someone should have cried, "Wait a minute!" One of these was the problem of synthetic detergents, to which phosphates were added in order to get laundry cleaner. When these phosphates flowed into the waters of rivers and lakes, they served as nutrients for plankton and algae. As a result, plankton multiplied in large numbers, and the waterways became choked by overgrowths of algae. In lakes, especially, fish died because the algae used up so much of the oxygen in the water. For that reason, state and local governments had begun to ban the sale of detergents containing phosphates. However, these laws were enacted, as Stans put it, "on a crazy-quilt basis geographically." He suggested that someone should say, "Wait a minute — what are we really doing here?" He then cautioned his audience:

Laws to ban phosphate detergents may give the public the false notion that the problem is being solved — but the fact is that nutrients,
including phosphates, will continue to pour into our lakes and rivers
from other sources, natural as well as man-made, and some of these
cannot be controlled. So if people assume a legal ban on phosphate
detergents will do the job, they may only lull themselves into neg-
lecting far more significant scientific efforts to help purify our waters
through new phosphate removal techniques in municipal waste treatment facilities.

Another example cited by Stans was the issue of selecting locations for new power plants. While the demand for electric power had kept growing, he said, it had become almost impossible to construct new generating plants, because of associated environmental considerations. Take the case of the city of Houston, he said, a city with "all the ingredients of growth — except enough electric power." Yet Houston was unable to build a new generating facility to meet its needs. Why? Because of the problem of the warm effluent water from the proposed plant. Even after costly cooling the effluent would still raise the temperature of the water into which it flowed — raise it, that is, a mere two degrees above the rigid temperature levels that had been prescribed in order to assure the preservation of marine life. "Isn't it time someone said: Wait a minute!" Stans asked, in view of the city's real needs for electrical power.

A third example was the question of banning the use of DDT. Since Ceylon had stopped using DDT, Stans pointed out, the number of mosquito-borne malaria cases had climbed to one million. Without DDT in India, he said, there would still be a 100 million cases of malaria each year, instead of only a few hundred thousand. Stans then brought up the controversy over the SST:

> For another example, Congress killed the SST. But shouldn't we as a nation have said, "Wait a minute. Are we so loth to build just two experimental models that we willingly sacrifice thousands of jobs, jeopardize the economic health of an entire city, forego the technological progress of a whole industry, and deny major benefits to our balance of payments?"

Stans went on to discuss other issues, including the delays in building the Trans-Alaska pipeline; the rigorous emission standards set for automobiles; and offshore oil drilling. He then offered several proposed guidelines:

> A determination of the economic impact should be required before environmental actions are mandated. The public must know what the cost will be, and whether it will get its money's worth.

He suggested that "panicky, *ad hoc*" approaches to problems of air, land, and water pollution must be avoided. Instead, feasible long-term plans to make improvements on a regular basis were needed. Stans concluded his address by saying:

> If we approach our problems in a spirit of fairness, we can meet our ecological needs and still satisfy our economic considerations, within the framework of continued technological progress.

About one month before Stans delivered his "Wait a Minute" address, he had sent President Nixon a memorandum in which he expressed similar views. This memo was headed "Costs of environmental compliance." Immediately, Russell Train, the chairman of the Council on Environmental Quality, submitted a memo of his own to President Nixon, headed "Environmental quality and economic progress." In his memo, Train wrote:

> This memorandum is in response to concerns expressed that environmental quality requirements will adversely affect the economy. I conclude that on the contrary, the economy will suffer little or no adjustment problems and that enhancement of the environment is consistent with economic development and growth.

Train then proceeded to support that contention in considerable detail. For example, referring to the changes of project delays resulting from environmental concerns, he said:

There is little doubt that environmental concerns have delayed some projects and even stopped a few. In assessing the extent and impact of these delays, both the particular facts and general economic factors must be considered.

Factually, many delays are unfairly blamed on citizen action to cover up poor planning. For example, environmental considerations have been blamed for widespread delays in power plant construction. In a recent Federal Power Commission survey of delays in 1970, however, only 2 out of 22 plants which were delayed up to 18 months were actually delayed because of environmental or other regulatory factors.

He proceeded to discuss investment by private firms, asserting that capital spending (plant investments, etc.) had declined during 1970 largely because of high interest rates, slow sales, and the general poor state of the economy, and that very little of the decline in capital spending was attributable to environmental reasons. Train then turned to a discussion of the expenditures required to improve the environment:

The costs of pollution control will grow significantly over the next five years, but their total impact on the economy should not be great. By 1975, the annual cost to meet air and water quality standards is only 7/10 of 1% of the Gross National Product (GNP).

The costs of meeting air and water pollution control standards for the most pollution intensive industries are generally less than 1% of the value of shipments and less than a 5% wage increase. . . .Any adverse trade impacts must be balanced against possibilities for expanding U.S. markets for pollution control equipment.

Train touched also on the economic damage caused by environmental degradation. When pollution worsens and the environment deteriorates, he said, illness increases, vegetation is harmed, and amenities are lost. Such losses far outweigh the costs of protecting and improving the environment. The costs to society from the damages of air pollution alone, said Train, had been estimated at more than $16 billion annually, over twelve times the amount spent on control equipment in 1970. He concluded that "at a time when other nations have begun to erode our technological lead in many areas, the pollution control technology being developed would provide a significant U.S. export market."

This debate between Stans and Train was then joined by Paul W. McCracken, the chairman of the Council of Economic Advisers. In a memorandum dated July 20, 1971, he submitted his own views to President Nixon. McCracken felt that Stans's positions contained a certain number of overstatements. When viewed in perspective, said McCracken, fears of economy-wide dislocations arising from expenditures on environmental protection and improvement appeared to be exaggerated. However, he wrote, "there will be a severe impact in some areas." He cited the automobile industry as the one that would feel perhaps the greatest impact:

It is not unreasonable to anticipate that car prices by 1975 will rise by roughly 15 percent more than the general price level due to pollution control requirements. Safety requirements will raise this to at least 20 percent. And pollution control devices may reduce efficiency and therefore add $75 or so to fuel costs per year per car. A relative price increase of this magnitude must be expected measurably to retard the growth of new car sales. While a number cannot be assigned to this with precision, an increase in auto prices of 20 percent relative to the price level generally could reduce the demand for new cars by roughly 1 million per year. This lower output might in part, of course, take the form of smaller cars, but either way substantial adverse effects on job opportunities in this industry are implied by pollution and safety requirements. While no other industry faces a similar cost burden, there will be a substantially above average impact on the costs of some others, including primary metals, paper and chemicals. Moreover, in these industries, unlike automobiles where costs will be imposed on domestic and foreign producers alike, some adverse effect on their international competitiveness can be expected.

McCracken then turned to the question of *costs* and *benefits*. It was true, he said, that (as Train had stated) total pollution control expenditures would not be excessively large, as a percentage of total gross national product — but the dollar amounts involved would be considerable. Something like $20 billion annually by 1975 had been predicted. Consequently, McCracken said, there was a need for "dispassionate consideration" as to whether the results obtained would be commensurate with the money spent. He then returned to the example of the automobile industry:

> Our policies have so far been formulated to treat many different problems separately with little attention to the priorities involved. The available evidence indicates, for example, that the air pollutants most clearly damaging to health and property are sulfur oxides and particulates. Yet the bulk of air pollution control expenditures will be incurred for automobiles, which produce neither of these pollutants but rather produce pollutants whose damaging effects are highly uncertain.

McCracken also discussed the pace and timing of spending on environmental problems. Haste, he pointed out, in taking policy actions often means wasteful spending. Expenditures would be more effective if made gradually. Long-term pollution control goals and productivity improvement could be more effectively attained, he suggested, by a more balanced and better-spaced approach. He went on to note that "the pollution control program has engendered some uncertainty whose resolution may be costly in certain cases," and mentioned automobiles once again:

> There appears, for example, to be considerable doubt that the automobile industry will be able to meet all requirements by 1976. Until

this doubt is resolved or the requirements changed, long-run planning
in this industry cannot confidently be undertaken. And change is
usually more costly when it is abrupt and unanticipated.

Just as McCracken had feared, America's automobile industry gradually began to decline. It turned out to be Japan, not America, that led the way in developing anti-pollution technology for automobiles. Moreover, when the first oil crisis began in the fall of 1973, and gasoline prices suddenly started climbing, Americans fell in love with small cars with good gas mileage. As we have seen, the result was 200,000 unemployed workers on the streets of Detroit and other auto cities. On the other hand, contrary to Train's expectations, pollution control equipment has not become a major American export. Pollution control technology is still only a peripheral sector, seemingly incapable of becoming a "mainstream" technology directly affecting production. When Secretary Stans called out "Wait a minute," perhaps Americans really ought to have paused a moment to cool their collective heads. Encouraged by President Nixon, many if not most Americans were so obsessed with environmental problems that they ended up killing a good many of the technological "geese that laid the golden eggs."

Nuclear Shutdown

The peaceful use of atomic energy was actively promoted in many countries around the world after President Eisenhower announced his "Atoms For Peace" program late in 1953. The public was told that a quantity of Uranium-235 the size of a baseball could produce as much heat as 177 railway carloads of coal, or 12,000 barrels of oil. There was widespread optimism about atomic power as the primary energy source for the next generation. In no time at all, the nations of the free world were riding the crest of an "atomic energy boom." Around 1963, however, some Americans came forward in opposition to the construction of nuclear power facilities. One of the very first such episodes involved a nuclear power plant that the Pacific Gas and Electric Company wanted to build at Bodega Bay, in California. In the June 22, 1963 issue, *The Nation* published a four-page article, entitled "Outrage on Bodega Head," that vividly described the dangers inherent in the proposed facility:

> The plant will contain 75 tons of 2½ percent enriched uranium fuel
> — 375 pounds of pure uranium, or about 150 times as much fission-
> able material as the Hiroshima atomic bomb. It will be kept from
> blowing up by a new safety system which has never been completely
> tested.
> The reactor will be about 333 yards — 1,000 feet — from the line
> of the San Andreas Fault, the same ever-active line of earth slippage
> that caused the 1906 San Francisco earthquake, as well as a few
> hundred lesser tremors before and since. (The one in 1906 turned
> over a locomotive just twenty miles from where the new reactor will
> be.)
> A 300-foot stack on the reactor plant will emit waste gases, in-
> cluding some charming stuff called Iodine-131, which is radioactive.
> I-131 decays fairly rapidly, but since the plant will operate all the

time, it will drop I-131 all the time. It will drop it, mostly, on Sonoma County dairy lands. This is an area which prides itself on the speed with which it gets milk to the market. It should be no trick at all for the Sonoma dairymen to get milk into California homes three or four days before half the I-131 has decayed.

The article included a map of the vicinity of Bodega Bay (Figure 12), which also showed the direction of the prevailing winds in the area. Anyone seeing this map would believe that Iodine-131 would be falling upon San Francisco. The article also declared that warm effluent (spent cooling water) from the plant would kill fish and shellfish, and that the plant itself would mar the lovely views around Bodega Bay. In fact, the slightly-enriched uranium (SEU) used in a nuclear power plant is incapable of causing a nuclear explosion like that of an atomic bomb. This fact was not well known then. To make a nuclear weapon, either more than 60-percent-enriched

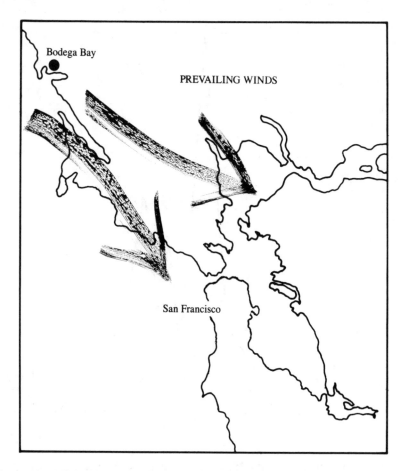

Figure 12   BODEGA BAY (CALIFORNIA) NUCLEAR POWER PLANT SITE

uranium, or plutonium, is required.

In 1964, Pacific Gas and Electric abandoned its plan to build the Bodega Bay plant. Opponents had scored their first victory without difficulty, but it was a localized, one-time affair. The campaign to oppose other nuclear power plants had not yet become an issue. The editor of *Science*, Philip H. Abelson, wrote in August 1968 that: "Then came a great outcry against air pollution associated with coal-fired plants. The move toward nuclear power became a stampede." Until 1968 there was a feeling that nuclear power plants were cleaner and better than coal-burning thermal power plants. This is indicated by Figure 13, which shows total annual orders for nuclear power reactors for 1953-1983 in the United States and Japan.

But starting in 1969, the atmosphere in the United States suddenly turned strongly anti-nuclear. Dr. Ernest Sternglass, a radiologist at the University of Pittsburgh, published an article in the *Bulletin of the Atomic Scientists* (April 1969) titled "Infant Mortality and Nuclear Tests," in which he maintained that the radioactive fallout

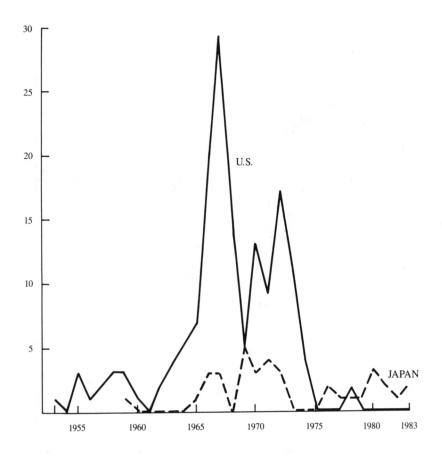

Figure 13   NUMBER OF ORDERS FOR NUCLEAR POWER STATIONS

produced by atmospheric nuclear weapons tests had caused increases in the number of miscarriages, stillbirths, neonatal deaths, and infant deaths. He declared that in the United States alone, 400,000 infants had died because of radioactive fallout. What Sternglass had done was to use statistical methods to investigate the relationship between nuclear weapons testing and the number of miscarriages and stillbirths — but most people were shocked by the figure of 400,000 infant deaths.

A number of other scientists attempted to refute the Sternglass thesis. Some pointed out errors in his statistics or in his use of the figures. Others noted that the radiation attributable to nuclear test fallout is equivalent to only about five percent of total natural level of background radiation, which was not enough to be fatal to fetuses or newborn infants. Sternglass, however, went on to say:

> Evidence that radiation in the form of nuclear fallout could have certain unanticipated severe side effects in the developing human embryo and infant should be a matter of deep concern. We should determine as surely as possible what the nature of such long-range effects might be before the country undertakes such potentially hazardous enterprises as digging a Panama Canal by setting off strings of hydrogen bombs amounting to as much as 150 megatons of energy, about one-third of the energy of all the bombs detonated in 17 years of atmospheric test-ing.

Quite naturally, in the minds of a great many people there came to be a link between nuclear power plants and "400,000 dead babies." Next, the notion became prevalent that nuclear generating plants are even worse than coal-burning ones. The August 29, 1969 issue of *Time* magazine noted that:

> Antinuclear critics argued that the damage to marine life from thermal pollution — excessive volumes of hot water discharged by nuclear plants — were far worse than the smog caused by smoke-belching power plants that used fossil fuels (oil and coal). They also voiced fears about possible harmful radiation effects.

In 1970, Arthur R. Tamplin and John W. Goffman published an even more influential study. They claimed that under the existing limits on nuclear radiation exposure (established by the U.S. government) the number of cases of cancer would soar dramatically, and that the maximum allowable dose should be cut to one-tenth of the safe level limits then in effect. Writing in the April 1970 issue of the magazine *Environment*, they argued that there was absolutely no need to rush headlong into generating electricity by means of a technology as dangerous as nuclear power. Then, on May 25, 1971, the U.S. Atomic Energy Commission announced the results of a surprisingly crude experimental safety test. The light-water type of nuclear reactor widely used at the time was equipped with an emergency core cooling system (ECCS) as a safety device. The ECCS was designed so that in the event of an accident, a large quantity of water would be dumped into the reactor to quickly cool down any nuclear fuel that might be in danger of overheating. However, when tests were carried out at the National Nuclear Reactor Test Station in Idaho, in a simulation it was found that the water would not flood the reactor. Afterwards the AEC discovered that this was because the simulator had been too small. In an actual reactor, the water would have indeed flooded the core. Nevertheless, the notion that nuclear reactors were dangerous became widespread — although, as can be seen from Figure 13, the anti-nuclear-power movement was not strong enough at the time to halt new orders for nuclear power plants.

In 1973, however, Ralph Nader turned his attention to nuclear power. After originally making a name for himself with his denunciations of the shortcomings of the automobile industry, he turned to additional consumer issues around 1969. Then, in early 1973, he announced that he had come to realize that the rush to develop nuclear power was one of the greatest mistakes in history. Later that same year, in an address to the Western Governors' Conference in Oregon, he said:

> Let me make a prediction here. I don't think that there will be another nuclear plant built in this country in nuclear fission after five years. I think there is going to be the biggest environmental, legal, legislative, executive branch, citizen, consumer battle in the history of the country.

On November 17-18, 1974, in Washington, D.C., Nader organized a conference, "Critical Mass 74," to halt nuclear power plant construction. Nader declared that nuclear power had to be stopped, and that solar power must be used as a safe energy source. At the time, he seemed to possess a magical ability to influence American thinking. In 1975, new orders for nuclear power plants fell to zero. Since then, eighty-seven plants already commissioned have been cancelled.

Nader cannot, however, claim complete credit for the severe downturn in the nuclear energy industry. Cold economics — escalating costs of construction combined with declining demand for electricity in the post-energy crisis era — is doing what thousands of hot-blooded demonstrators never could, claims Christopher Flavin. It is slowly closing down the nuclear industry.

The Japanese tendency to imitate Americans was evident again. In nuclear power, Japan continued to emulate the United States until very recently. When President Eisenhower announced his "Atoms For Peace" program in 1953, Japan picked up the cue and began research on peaceful utilization of atomic energy. At that time, even the progressive Socialist Party in Japan supported nuclear power. The April 1957 issue of the party's official magazine published the essay "Trends in Peaceful Uses of Atomic Energy," by Shigeyoshi Matsumae, a Socialist member of Parliament. He wrote: "The Japan Socialist Party is continuing its active efforts on behalf of research and development to encourage peaceful utilization of atomic energy." Nevertheless, when the American movement against nuclear power began to emerge, Japan's Socialist Party scrapped its former policy and began giving support to anti-nuclear-power campaigns organized by local residents in areas where nuclear plants were planned. The Socialist Party organ, in its May through August 1970 issues, ran a series of articles by Kei Inoue, under the title "The Advance of Nuclear Power, and Its Perils: The Dangers of Radioactive Pollution." Later, in the May through August 1972 issues of the same magazine, there appeared a series signed by "the Japan Socialist Party Central Headquarters Committee to Support the Movement Opposing Nuclear Power."

In July 1973, the Japan Congress Against Atomic and Hydrogen Bombs (Gensuikin) invited Dr. Arthur Tamplin, the father of the anti-nuclear-power movement in

America, to visit Japan. In a news story headlined "New Impetus For Citizen Campaigns From Dr. Tamplin," the Tokyo newspaper *Mainichi Shimbun* wrote:

> It had been hoped that the invitation to Dr. Tamplin would serve as a shot in the arm to stimulate the anti-nuclear-power movement, and indeed, Dr. Tamplin's five days of energetic activity have amply fulfilled those hopes.

The Japanese media and opposition political parties serve as watchdogs against ruling party abuses, much as consumer groups do in the United States regarding the political party in power. The uproar over Japan's first and only nuclear-powered ship, the *Mutsu*, clearly demonstrates this sudden shift in the attitude. In November 1967, it was decided that the vessel's home port would be Ominato Harbor in the city of Mutsu, at the northern tip of Japan's main island of Honshu. At the time, the local newspapers unanimously hailed the ship as a godsend, and proclaimed that the nuclear age had now indeed arrived. But as the construction of the nuclear-powered vessel proceeded (1972-73) and its "nuclear fires" were finally going to be lighted, the opposition movement suddenly gained strength and took on a new militancy. Because of protests from fishermen fearing water contamination, it proved impossible to conduct trials of the ship's nuclear reactor in Ominato Harbor. The government decided to run the tests out at sea. The vessel was supposed to leave port on August 27, 1974, but at dawn that day, 210 small fishing vessels totally surrounded the *Mutsu*, blocking her from sailing out of the harbor.

After seven o'clock that evening, as darkness fell, a brisk wind sprang up, threatening the fleet of fishing boats and forcing them to leave the side of the *Mutsu*. At last the ship was able to get underway. But when the *Mutsu* finally reached the high seas under auxiliary power, and the nuclear reactor was turned on, the reactor's shielding was discovered to be defective, allowing radiation to leak toward the deck. Like the beam of a flashlight shining through a hole in a curtain, a beam of radiation was emitted from a crack in the reactor shield. The nuclear reactor was shut down. The radiation exposure resulting from the leak was extremely small: 500 hours of continuous exposure to this beam would be needed to equal the radiation received from a single ordinary chest X-ray. Nevertheless, fishermen living in or near the *Mutsu's* home port gained the impression that the nuclear ship was discharging radioactive materials, and this time they unfurled a "keep the *Mutsu* out of the harbor" campaign. The vessel was unable to return to Ominato Harbor, and was forced to drift at sea for several days. That happened more than ten years ago. To this day, the *Mutsu* has never actually operated under nuclear power.

Techno-Skeptics

Crusades usually begin with good slogans. In 1973, a German-born British economist, the late E. F. Schumacher, came up with a handy and attractive catchphrase: "Small is beautiful." In his book with that title, Schumacher declared that "giantism" is a leftover from the nineteenth century, and that technology from now on must be small:

> I have no doubt that it is possible to give a new direction to technological development, a direction that shall lead it back to the real

needs of man, and that also means: *to the actual size of man*. Man is small, and, therefore, small is beautiful. To go for giantism is to go for self-destruction.

Schumacher described this "human-sized technology" in these words:

What is it that we really require from the scientists and technologists? I should answer: We need methods and equipment which are:

— cheap enough so that they are accessible to virtually everyone;
— suitable for small-scale application; and
— compatible with man's need for creativity.

He also referred to this human-scale technology as "intermediate technology." Schumacher himself was the founder and chairman of the Intermediate Technology Development Group in Britain, which devises small-scale tools and machinery for use in developing countries. Schumacher was not necessarily opposed to all technology, but did object to large-scale technology. But as can be seen, for example, from the fact that the anti-science advocate Theodore Roszak wrote the preface to *Small Is Beautiful*, Schumacher ultimately represented a basically hostile attitude toward science and technology. He was also against nuclear power:

Of all the changes introduced by man into the household of nature, large-scale nuclear fission is undoubtedly the most dangerous and profound. As a result, ionising radiation has become the most serious agent of pollution of the environment and the greatest threat to man's survival on earth.
The continuation of scientific advance in the direction of ever increasing violence, culminating in nuclear fission and moving on to nuclear fusion, is a prospect of terror threatening the abolition of man.

Schumacher's catch-phrase "small is beautiful" had a poetic ring to it that attracted artists and intellectuals. It became a popular phrase in Japan, and played a major role in impeding certain research and development of large-scale technologies. Today, however, the Japanese popular phrase "small, flat, thin, and narrow" would make him happy.

In the wake of "small is beautiful," catch-phrases like "hard path" and "soft path" were coined in 1976 by an American-born British physicist, Dr. Amory B. Lovins, in an article entitled "Energy Strategy: The Road Not Taken," which appeared in the October 1976 issue of *Foreign Affairs*. Lovins referred to heavy-industry, centralized energy systems — such as nuclear generating plants and coal-fired thermal electric power plants — as the "hard path," in contrast to regenerative energy systems like solar and wind generators or small hydroelectric generators, which were labeled the "soft path." If soft energy were utilized, he claimed, there would be no need for dangerous "hard" types of energy, like nuclear power. Lovins expanded this thesis in the book, *Soft Energy Paths: Toward a Durable Peace*, which was translated into Japanese in 1979. Lovins visited Japan in April 1980, giving lectures and participating in discussions all over the country. At that time, he asserted that if housing insulation and heat exchangers were used, there would be no need for heating and air conditioning, which would mean great savings in energy demand.

Lovins recommended wind-power generators for Japan, saying that those with a power output of up to 100 kilowatts were most appropriate. According to the calculations of experts, a 100-kilowatt wind power generator needs to have a propeller nearly 100 feet in diameter. The tower supporting that propeller must be tall enough to reach winds strong enough to turn the propeller, so a tower height of at least 165 feet is required. Moreover, in order to obtain a full 100 kilowatts from one such generator, a strong wind is needed. On a calm day, or one with a light breeze, power output would be zero. Clearly, there are practical constraints on this type of energy for industrial uses.

The Tokyo office of the *Asahi Shimbun* uses 2,700 kilowatts of power to run its presses when printing the newspaper. To supply this much electricity using 100-kilowatt wind-powered generators alone, 27 such generators would have to be placed on the roof of the building. That roof, however, is only 300 feet long by 200 feet wide. Each generator would require a circular space at least 100 feet in diameter, so that the direction in which the propeller faced could be changed according to the wind direction. Consequently, only six wind-powered generators could be placed on the roof of the building. Where would the other 21 go? Compared with nuclear and coal generated power, wind and solar power are extremely low-density energy sources. They are weak, diffused energy sources, and to harness them in such a way as to provide the needed quantities of electricity is very difficult, both from a theoretical and a practical or technological point of view.

When I described Ivan Illich's "bicycle theory" in Chapter II, I suggested that he should have tried traveling to Japan by bicycle rather than by jet aircraft. Similarly, I feel, Lovins ought not to have flown to Japan in a jet that burned fuel made from petroleum. If he really believed in wind power, he should have come to Japan on a glider or in a sailboat.

The reader might get the impression that I am opposed to research and development on technology for utilizing solar energy and wind power. That is not the case. I believe that full encouragement must be given to research and development in solar energy, wind power, wave power, tidal power, geothermal power, and all other forms of alternate energy. But I also believe that those energy sources are not going to be able to run railways or operate huge factories, just as a glider will never offer an adequate substitute for a jet transport aircraft.

Saburo Okita, an economist and former Japanese foreign minister, contributed a foreword to the Japanese-language edition of *Soft Energy Paths*, in which he wrote:

> I personally do not necessarily agree with everything advocated by this book. I believe that in a country like Japan, with an inadequate supply of energy in absolute terms, nuclear power generation is a necessity. Moreover, I do not consider the "soft energy path" and the "hard energy path" to be mutually exclusive. Rather, they are mutually complementary.

As a result of these anti-scientific views in America, the development of the supersonic transport was halted, the automobile industry was thrown into decline, and plans for nuclear power plant construction were torn up. Aircraft, automobiles, and nuclear reactors were three of America's key basic industries. The effects of the anti-science disease upon the technology and the economy of the United States was significant.

In Japan, too, the anti-science disease can be considered as hindering the development of various technologies. A typical case involves the development of proteins manufactured from petroleum, generally known as "petro-proteins." The process utilizes paraffin as a raw material, which is present in crude oil, to produce protein substances that can be used in cattle feed and in nutrients for fish farming (aquaculture). Specifically, yeasts or similar microorganisms are placed in a mixture of normal paraffin and water. The yeasts feed on the paraffin and multiply. Since the cells of the yeast organisms are composed of protein substances, they can be harvested and processed to make an excellent protein nutrient. This petro-protein manufacturing technique began to be discussed in the Japanese press in the autumn of 1968. Newspaper accounts at the time reported that if wheat flour were mixed to contain four percent petro-proteins, the flour would have a nutritional value equal to egg whites.

Microorganisms like yeast multiply with amazing rapidity. For example, a single yeast cell will divide once every 20 minutes, becoming two cells, then four cells in another half hour, then eight cells, and so on. In 24 hours, therefore, the number of cells would be almost inconceivably large — something on the order of 30 *billion billion* cells.

At the present time, soybean cake and fish meal are used to supply a protein component in animal feed. When soybeans are grown in fields, however, each single soybean sown will yield only several hundred soybeans, six months later. Thus, from the standpoint of protein production, the soybean can be said to grow at a snail's pace, while yeast multiplies at nearly the speed of light. Accordingly, if petro-proteins could be put to practical use, experts predicted, the world's whole food situation might be drastically improved. Some of the Japanese manufacturers doing research and development on the process hoped that someday petro-proteins would be used not only as a nutritional "extender" for sausages and as a base for making artificial meat, but also as a raw material for protein fibers and synthetic leather. In the fall of 1969, however, the safety of petro-proteins became an issue. On October 24, 1969, the Resources Council of the Japanese government's Science and Technology Agency issued a report that concluded:

> A microbial industry utilizing petroleum is needed; but the safety of its products has not yet been adequately corroborated, and caution is called for. The Japanese government must take systematic measures, from a comprehensive standpoint, to ensure that research and development maintain adequate safety.

The report indicated three questions that required answers: was there any danger of impurities getting into the normal paraffin used as the culture medium? Could the manufactured proteins possibly be toxic to animal life? And finally, were livestock and fish raised on petro-proteins really safe for human beings to eat? Independent studies of these questions were begun by three agencies: the Microbial Protein Feed Research Council, an advisory body of the Ministry of Agriculture; the Protein Committee of the Food Sanitation Board, an advisory organ of the Ministry of Health and Welfare; and the Petro-Protein Committee of the Light Industry Productivity Council of the Ministry of International Trade and Industry (MITI). During 1969, MITI announced its conclusion that petro-proteins were harmless, and in July 1971, the Agriculture Ministry issued a report to the effect that the proteins were safe as long as the individual companies took adequate precautions when processing them into animal feed. The Health Ministry's study took much more time, but it concluded on December 15, 1972, that petro-proteins were safe to use as animal feed.

Around that time, however, consumer groups put pressure on the Health Ministry, declaring that they opposed the licensing of petro-proteins as animal feed. By 1972, the anti-science disease was rampant in Japan. Simultaneously, activists from America turned up in Japan, and the following appeared in one of the Tokyo newspapers at the time:

> Mr. Jack Tropp, the Japanese liaison officer for an American organic agriculture group called the Unpolluted Food Movement, spoke with impatience in his voice as he handed over to a consumer group some information on petro-proteins that have not yet been licensed in the United States. "In America," he admonished the Japanese consumers, "we would organize a boycott movement, and not eat any meat from animals fed on petro-proteins. At least," he went on, "we wouldn't back down until we had a disinterested third party perform much more rigorous safety tests."

By the beginning of 1973, housewives in Tokyo demanded that the Minister of Health forbid the manufacture, sale, or use of petro-proteins. Among the reasons they cited were these:

> Petro-proteins are a totally unfamiliar substance, with which human beings have had no previous experience in their eating habits. Despite that fact, the safety of these substances has been judged merely on the basis of studies of the data provided by the manufacturers themselves. This kind of procedure represents a violation of the consumer's "right to demand safety." Even if these substances are used solely to feed livestock and fish, there is still a danger that harmful impurities will ultimately accumulate in the human body.
>
> The manufacturers have not necessarily released all of their safety test data to the public. Even the names of the species of microorganisms used in the manufacturing process are company secrets, and have not been revealed. That is an infringement upon our "right to know."
>
> Natural animal protein supplies have been reduced as a result of pollution and other causes. For the government to fail to take measures to deal with those problems, while instead forcing consumers to eat meat, milk, and eggs from animals fed on petro-proteins, is a violation of the consumer's "right to choose."

The principal "harmful impurity" at issue was 3,4-benzopyrene, a notorious carcinogen, which was suspected of being present, in amounts of less than one part per billion (ppb), in the normal paraffin used to grow petro-proteins. Housewives reflecting consumer sentiments panicked at this, without stopping to consider how minute an amount of the chemical might be involved. They emotionally proclaimed their total and absolute opposition to the proteins. The very word "cancer" has that kind of effect upon people.

However, 3,4-benzopyrene is often found in natural foodstuffs: for example, 2 to 37 ppb in smoked fish, 7.4 ppb in spinach, and 3.9 to 21.3 ppb in tea. Table 1 shows the amounts of this chemical occurring naturally in various foods. If 1 ppb or less of 3,4-benzopyrene were not permissible in petro-proteins, then no one should be allowed to eat sausage or cabbage, or drink coffee. But the housewives who raised outcries of opposition totally ignored the whole matter of quantities. Opposition to quantification is one of the classic symptoms of the anti-science disease, and this episode was an excellent illustration of that. A group of housewives stormed into the Ministry of Health and Welfare and for four hours grilled the section chiefs who were handling the petro-protein issue. The tenor of their protest, as reported prominently in the press and on television, was something like this:

> All these years, we consumers have been deceived and fooled. The officials just tell us wait — wait until petro-proteins are actually used in animal feed. Afterwards they'll tell us they're so sorry about the harmful effects, and offer to pay compensation. But they won't be able to make up for the damage that easily by paying money!

Table 1　3,4-BENZOPYRENE CONTAINED IN VARIOUS FOODS
(figures represent parts per billion)

Food	Minimum	Maximum
Broiled meat	0.17	0.63
Charcoal-broiled meat	2.60	11.20
Steak		50.40
Ham, sausage	0.02	14.60
Smoked meat	23.00	107.00
Smoked fish	2.10	37.00
Gas-broiled fish		0.90
Spinach		7.40
Tomatoes		0.20
Cabbage	12.60	48.10
Soybeans		3.10
Apples	0.10	0.50
Other fruit	2.00	8.00
Saturated vegetable oil	0.40	36.00
Margarine	0.20	6.80
Coffee	0.10	4.00
Tea	3.90	21.30
Baker's yeast	1.80	40.40
Whiskey		0.04

Based on an analysis by the World Health Organization (WHO)

One of the companies that had already started to build factories, intending to manufacture petro-proteins, announced on February 20, 1973, that it would postpone commercial production until such time as the public was prepared to accept the proteins. A spokesman for the firm stated:

> We have both a conviction and scientific proof that petro-proteins are safe. Users of animal feed have been urging us to begin commercial production as soon as possible. Production would be right, from the standpoint of the country's best interests, but since there is little likelihood of public acceptance, the reputation of our company would suffer if we simply tried to force this upon society. We have decided, therefore, to postpone indefinitely the commercial production of petro-proteins, until there is a public consensus and we are given administrative guidance by the Ministry of Agriculture.

The next day, February 21, five other companies also made known their decision not to commercialize petro-proteins, and to give up all plans for producing them. Said the president of one of the firms: "Our change of heart wasn't solely a result of the housewives' movement against petro-proteins. If we try to go against public opinion, we won't be able to stay in business." A promising new technology was nipped in the bud.

In 1975, a similar controversy broke out in Japan. A campaign developed to oppose the practise of adding a substance called "lysine" to bread served in school lunch programs. Lysine is one of the so-called essential amino acids that build up the protein molecules of which the human body is largely composed. Some of these amino acids are synthesized naturally within the body, but others are hardly synthesized, or not at all. Those amino acids not synthesized must be obtained in food, or else the body cannot remain healthy. The problem has to do with the degree to which each of the essential amino acids actually occurs in various foods. Human milk, considered a high-quality protein source, contains ten essential amino acids in wellbalanced proportions. In other words, the essential amino acids in human milk can all be fully utilized by the body as the raw materials for building proteins.

In the case of wheat flour, however, the amounts of the various essential amino acids are unbalanced, with lysine particularly lacking. Excess amounts of other amino acids cannot be used by the body to make proteins, and are virtually useless. For that reason, scientists had the idea of adding a small amount of lysine to wheat flour, in order to correct the imbalance of essential amino acids. The lysine to be added is produced in large quantities by a special variety of bacteria, a mutation developed by exposing ordinary bacteria to radiation or ultraviolet light. When properly cultivated, these bacteria multiply with amazing rapidity, releasing large amounts of lysine in the process.

When one group of primary school children was given bread containing lysine, while another similar group was given bread without lysine, and the body growth rates of the members of the two groups were measured, the children who had eaten the lysine bread were found to have grown nearly one inch more than the non-lysine group, on the average, in the course of only one year. In view of those results, many local governments in Japan began adding lysine to the bread in their school lunch programs, beginning in 1968. In the spring of 1970, however, when the Tokyo

94

metropolitan government decided not to add lysine to school lunches, a controversy erupted. The opposition to lysine was based on a belief that man-made lysine contained the carcinogen 3,4-benzopyrene. An analysis found the latter chemical in amounts of from 0.06 to 0.30 parts per billion in lysine.

Looking at Table 1, it is obvious that ordinary daily foods and beverages contain far more 3,4-benzopyrene than is found in lysine. Thus, once again, any considerations of quantity were totally ignored. Yataro Tajima, the director of Japan's National Genetics Institute, wrote the following in his book *How Does the Environment Influence Genes?*, published in 1981:

> It seems to me that when the amounts [of 3,4-benzopyrene] involved are so minute, any debate over the advisability of adding lysine must inevitably move beyond the boundaries of science, into something more like faith or religion. One has the feeling that scientific judgment is absent from the argument.

By 1975, most people in Japan had come to question science and its benefits. An emotional outcry by the mothers succeeded in ending the use of lysine in school lunches. Today, even with the boom in biotechnology, the average Japanese is somewhat distrustful (and rightly so) about a number of new products that contain chemicals that could have a long-term negative impact on one's general health.

VI
THE ANTI-SCIENCE CULT

Man has always respected and feared the unknown or mysterious. The basic human instinct is still the same: human beings have irrational responses to what remains unexplained.

In Washington, D.C., on March 30, 1981, President Reagan was shot and wounded by a pistol in the hands of a man in his early 20's. It was Monday, slightly chilly, with a sprinkling of rain. The President had just delivered an address to the annual convention of the construction trade unions at the Washington Hilton Hotel, and was leaving the hotel through a VIP exit. A crowd of people hoping to catch a glimpse of the President had gathered, and a group of reporters and television cameramen were also standing in a designated area a few yards from the door through which the President and his party emerged. The President, as he always does, smiled and waved to the crowd of onlookers. At that instant, shots rang out. A man standing in the group of journalists had fired a pistol. A police officer fell, then a Secret Service agent, then the President's press secretary. Another Secret Service agent shoved the President into his black limousine, which sped away from the scene.

The incident occurred at 2:26 P.M., and I and the others at the Woodrow Wilson International Center for Scholars learned about it almost immediately. A television set was temporarily placed in the Center's spacious library, which also serves as a conference room, and most of the Fellows of the Center gathered there to watch the live news reports.

The assailant had fired his pistol just after the television cameramen had started recording the scene on videotape. The cameras had thus preserved the sight of the police officer turning around slightly as he was shot, and the startled expression on President Reagan's face. The unforgettable scene was replayed over and over again on television. One of the Wilson Center scholars watching the broadcast remarked: "In this country, you never know when you're going to be shot. Even the President is no exception." About a month before the attempted assassination, that particular scholar had been held up and robbed just outside the apartment building in Washington, D.C. where he and I both happened to be living during our terms at the Center. After the robbing, he had written the following warning memo and sent copies to all the other residents of the apartment building:

> This is simply to advise you of a fact of which you are probably already aware: BE CAREFUL IN THIS VICINITY AT NIGHT. Last

night, Tuesday, March 3, I was robbed at "broken bottle-point" just outside the south door of the ———— Apartments in the well-lit covered parking area. I was unhurt, but I did give him what little money I had on me. The moral I learned: do not do my grocery shopping at Safeway at 10 at night. As you know, there is no outdoor security for these buildings. And the police are undoubtedly busy elsewhere in the city. So, take it from one victim: be extremely careful in your neighborhood at night. Do not go out unless you absolutely have to. From now on, I won't.

In Japan, ownership of pistols and other weapons is rigorously controlled by law, but in America, states have few restrictions on the purchase and ownership of guns. According to statistics, two million firearms are sold annually in America, where 45 percent of all households have guns. The number of murder victims in 1979 was more than 21,000, and it was estimated that the figure would grow by 10 percent in 1980. On the Sunday after the shooting of President Reagan, the *Washington Post* carried a column by the paper's astrologer, Svetlana Godillo. After mentioning her initial shock at learning of the attempt on the President's life, she said:

And yet, as an astrologer, I breathed a sigh of relief, for the "other shoe had fallen." The ominous aspect that was dogging the charts of Ronald and Nancy Reagan, accentuating the link between the president and the vice president, the ominous aspect that appeared on the charts of the Inauguration and of the eclipse of Feb. 4 (that fell on Reagan's Sun) had materialized — once such an aspect materializes, it usually does not reoccur. And if astrology worked, President Reagan was going to be okay.

Astrology, "invented" several thousand years ago in ancient Mesopotamia, was concerned with the destinies of kings and emperors, courts and royal families all controlled by the movements of the planets. Gradually, however, astrologers came to believe that if the planets could control the fates of kingdoms, then surely they must also govern the destinies of ordinary individuals. This type of personal astrology developed more than 2,000 years ago, and spread throughout the ancient Greek and Roman worlds.

In astrology the position of the planets in the zodiac at the moment of a person's birth has a major influence upon that individual's future. Astrology has no scientific basis whatsoever. Dr. Carl Sagan, the astronomer, made this comment in his best-selling book, *Cosmos,* that I translated into Japanese:

Astrology can be tested by the lives of twins. There are many cases in which one twin is killed in childhood, in a riding accident, say, or is struck by lightning, while the other lives to a prosperous old age. Each was born in precisely the same place and within minutes of the other. Exactly the same planets were rising at their births. If astrology were valid, how could two such twins have such profoundly different fates?

Astrology is childish, unscientific stuff, but from about 1968, when America's anti-science disease started to become serious, astrology took the public by storm. During 1969 and 1970, a number of American magazines devoted special attention to the craze for astrology. For example, the weekly magazine *Time*, in its issue of March 21, 1969, ran a six-page story on the astrology phenomenon that included vignettes like this:

> In the basement of the Shambala Bookstore on Berkeley's Telegraph Avenue near the university's campus, 20-year-old Sheila O'Neil looked up from her calculations on the chart before her and shook her head. "We'd better postpone the organization meeting until next week," she said. "Mercury's going into opposition with Saturn in the 3rd House, which will mean bad communicating. But next Tuesday all systems will be go."

According to the account in *Time*, bookstores specializing in astrological works were patronized even by graduate students and instructors. At the time, there were said to be 10,000 professional astrologers in America, and 17,500 part-time practitioners. There were even companies that offered computerized astrological predictions. Department stores sold cocktail glassware and other articles decorated with the scorpion, the water bearer, the fishes, the crab, the lion, and the other symbols of the zodiac. Perfume companies concocted new scents associated with the twelve astrological "houses," and rang up big sales figures. In fact, 1,200 of America's 1,570 newspapers were reported to be carrying regular astrology columns. It was as if even the country's newspaper editors had come down with a bad case of the anti-science disease. And not only editors, but many reporters as well, appeared to have fallen victim to the epidemic. These newsmen repeatedly questioned California's Governor Ronald Reagan as to whether he was using astrology to govern the state. Replied Reagan: "I'm no more interested in astrology than any average person would be."

The American vogue for astrology was added to Japanese superstitions. In Japan, women's weeklies and light entertainment magazines began to print articles and features on astrology, and continue to do so today. But it would be unthinkable for any Japanese newspaper, proud of its objective coverage of events, to carry astrological columns or articles. Referring to the astrology columns in American newspapers, Carl Sagan wonders, "why are they published as unapologetically as sports statistics and stock market reports?"

James Michener, the best-selling novelist, commented:

> I think it's rather shocking the way the media has surrendered to this, so that major newspapers will devote almost a whole page to the addlepated guesses of someone who has given eighty-eight predictions. One turns out to be almost true, and that proves clairvoyance.

Astrology started becoming fashionable, I believe, as the anti-science disease worsened, and people lost their faith in rationalism and the scientific way of thinking.

Along with astrology, Americans even revived interest in the *I Ching*, a book used by the ancient Chinese for divination. The actual contents of the work itself comprise a kind of wisdom about life. But to try to use the book as a device for making predictions about the future is utterly without scientific basis. Says James Michener:

> As I worked through the youth movement in those years, I was shocked to find that an enormous number of them were addicted to the I-Ching nonsense, coming out of China, in which it was assumed that a book of recondite material assembled 1,800 years ago was able to produce solutions to current problems.

Michener's criticisms notwithstanding, many Americans, even today, are still fascinated by unscientific fads like astrology or the *I Ching*.

The UFO Craze

Another symptom of the anti-science disease was a revival of the craze over Unidentified Flying Objects (UFO's). By 1966, a full-fledged UFO "boom" was underway. By counting the number of articles about UFO's in the *New York Times* during a given year, some notion can be gained of the scope of the UFO craze. Figure 14 shows this in graphic form. During the early 1960's, the *Times* carried five articles

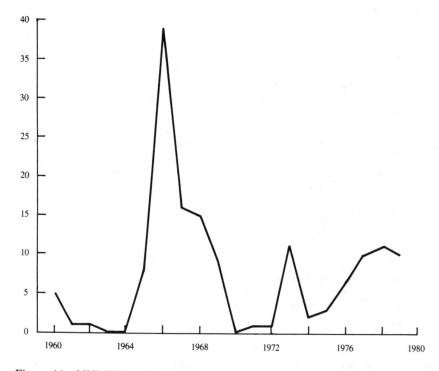

Figure 14   NUMBER OF ARTICLES ON UFO's IN THE *NEW YORK TIMES*

per year, at the most, and none at all in 1963 and 1964. But in 1965, the number jumped to eight, and soared to a total of 39 articles in 1966. The boom tapered off until about 1969 and then disappeared. Then came a small peak in 1973, and yet another peak in the late 1970's.

UFO's (commonly called "flying saucers" or "flying discs" until the 1950's) first appeared in the news in 1947. On June 24, Kenneth Arnold, of Boise, Idaho, was piloting a small private aircraft from Chehalis, Washington, to Yakima, Washington, when he suddenly saw a bright flash reflected on his wing. Wondering what it was, he looked around and saw what appeared as a line of nine peculiarly-shaped aircraft flying toward Mt. Rainier. "I could see their outline quite plainly against the snow as they approached the mountain," Arnold later reported. "They flew very close to the mountain tops, directly south to southeast down the hog's back of the range, flying like geese in a diagonal chain-like line, as if they were linked together."

According to his description, the objects seemed smaller than a DC-4 airliner, and unlike any conventional aircraft: they were "saucer-like" and "flat like a pie pan and so shiny they reflected the sun like a mirror." He estimated their speed at about 1,200 miles per hour. The newspapers played up Arnold's story, but with a tone of amusement and disbelief. Angered, Arnold retorted that if he ever again saw a phenomenon in the sky — "even a ten-story building flying through the air" — he would not say a word about it. In the days and months that followed Arnold's experience, more and more observers came forward claiming to have seen "flying saucers" or other such objects. Only a few days after Arnold's sighting, a disc was reported seen over his hometown of Boise — "half-circle in shape, clinging to a cloud and just as bright and silvery looking as a mirror caught in the rays of the sun."

Stories about "flying saucer" sightings began to snowball. At Muroc Air Base, California, a group of Air Force officers reported seeing disklike objects whirling through the sky at a speed in excess of 300 miles per hour. In Portland, Oregon, several policemen told investigators they had seen a group of discs that "wobbled, disappeared, and reappeared" several times. A series of "flying saucer" reports also came from pilots of commercial and military aircraft. On January 7, 1948, an unidentified object that resembled "an ice cream cone topped with red" was sighted over Godman Air Force Base, in Ft. Knox, Kentucky, by several military and civilian observers. Air Force Captain Thomas F. Mantell and three other Air National Guard F-51 pilots were ordered to investigate the phenomenon. Three of the planes, including Mantell's, closed in on the object, and reported that it was "metallic," of "tremendous size," "round like a teardrop," and "at times almost fluid."

Captain Mantell contacted the Godman control tower with an initial report that the object was travelling at half his speed at "12 o'clock high," or directly ahead of and above his own aircraft. "I'm closing in now to take a good look," he radioed back. "It's directly ahead of me and still moving at about half my speed. . . .The thing looks metallic and of tremendous size." Then he reported: "It's going up now and forward as fast as I am — that's 260 miles per hour. I'm going up to 20,000 feet and if I'm no closer, I'll abandon chase." Those were the last words ever heard from Mantell, at 3:15 P.M. Later that day, his body was found in the wreckage of his plane near Ft. Knox. Subsequent investigation revealed that Mantell had probably blacked out at 20,000 feet from lack of oxygen, and had died of asphyxiation before the crash. The Air Force initially announced that the mysterious object that Mantell had chased to his death was probably the planet Venus, which is very bright and is frequently visible by daylight. However, further investigation disclosed that Venus would not have been visible at that time in the reported positions of the mysterious object. Thus, the riddle of what Mantell was pursuing remains unsolved.

As reports of "flying saucer" sightings continued in a never-ending stream, the U.S. Air Force, on January 22, 1948, finally established a special team at Wright-Patterson Air Force Base, at Dayton, Ohio, to investigate the reports. The United States and the Soviet Union were then in the midst of the "cold war." If the "flying saucers" should turn out to be the latest Soviet secret weapon, that would have serious implications for the national defense of the United States. The Air Force at least needed to find out whether the "flying saucers" were of Russian origin. The investigation continued for more than one year, and the results were announced on April 27, 1949. According to the Air Force's report, most of the so-called "flying saucers" were definitely shown to have been meteorological balloons, distant jet aircraft, rockets being tested, flocks of birds, falling meteors, optical illusions, or cases of mistaken identification. In some cases, false reports and fabrications were involved. There was no evidence that any of the objects were of Soviet origin nor were they spaceships from beyond our planet.

On April 30, 1964, the U.S. Air Force released the results of 16 years of investigating unidentified flying objects. According to that report, there had been 812 reported sightings of UFO's during that 16-year period, but 92.3 percent of these cases, far from being "flying" objects, actually involved visual misinterpretations of what had been seen, or mistaking weather balloons and similar objects for "flying saucers." The remaining 7.7 percent of the reports could not be definitely resolved, but in view of the short periods of observation and the lack of clear or detailed descriptions of what was seen, the objects sighted were not considered to pose any threat to America's national security. With the publication of the Air Force report, the UFO phenomenon died down for the time being.

The 1966 outbreak of the UFO craze began with a report that four spaceships had landed in a swamp in Ann Arbor, Michigan, on March 22, 1966. Eyewitnesses reported that the spaceships had landed and then taken off, repeating this action three times. Professor Joseph A. Hynek, of the Northwestern University Department of Astrophysics, declared that in his opinion the "spaceships" had actually been methane, or "swamp gas," wisps of which had ignited and floated above the swamp. In January, 1969, a team of researchers at the University of Colorado disclosed the results of an 18-month study. The scientists asserted that there was absolutely no evidence that UFO's were spacecraft from elsewhere in the universe, and that further investigation along those lines would be a waste of time and money. Such assertions, however, do not calm the popular fears about the subject.

Sometimes, people see something in the sky and cannot tell what it is, and immediately leap to the conclusion that it must be a spaceship from some other world. To such individuals, "UFO" is synonymous with "spacecraft from another galaxy." No matter how many scientists investigated UFO's from whatever possible angle and no matter how often it was determined that no extraterrestrial spacecraft were involved, the number of people who absolutely refused to believe the scientists continued to increase.

Even some American politicians were affected by this trend. A *New York Times* reporter on September 14, 1973, quoted Jimmy Carter, then serving as governor of Georgia, as saying that he had recently seen a UFO, and that he had seen one on another occasion prior to becoming governor. Moreover, in November, 1977, after becoming President, he asked NASA to look into the matter of the UFO's. At that time, Dave Williamson, NASA's assistant director for special projects, commented:

> It's not wise to do research on something that is not a measurable phenomenon. There is no measurable UFO evidence such as a piece of metal, flesh or cloth. We don't even have any radio signals. A photograph is not a measurement.

When pressed, NASA finally refused President Carter's request. NASA's head, Robert Frosch, in a letter to the presidential science adviser, Dr. Frank Press, wrote as follows:

> Such an inquiry would be wasteful and probably unproductive. The National Aeronautics and Space Administration stands ready to analyze any bona fide physical evidence from credible sources, but such evidence has never been found.

Dr. Carl Sagan wrote the following in his book *Cosmos*:

> The critical issue is the quality of the purported evidence rigorously and skeptically scrutinized — not what sounds plausible, not the un-substantiated testimony of one or two self-professed eyewitnesses. By this standard there are no compelling cases of extraterrestrial visi-tation.

With the deepening of the distrust in science, increased attention was paid to a variety of pseudo-scientific subjects. From the late 1960's into the 1970's, a popular topic for speculation, one that rivalled the UFO, was the "USO," or "unidentified swimming object." The most famous USO is the aquatic monster nicknamed "Nessie," which is supposed to dwell in the depths of Loch Ness in Scotland. Loch Ness, a long, narrow lake about 23 miles long and generally about one mile across, has a maximum depth of 770 feet. Its reddish-brown waters fill the Great Glen, the rift-like valley that cuts east and west through the mountains of Scotland, bisecting the country. The tradition that a strange creature lives in the waters of Loch Ness goes back many centuries. More than 1,300 years ago, St. Adamnan, a cleric and scholar of the northern British Isles, wrote the following account in one of his Latin works:

> In about the Year of Our Lord 565, the blessed St. Columba of Ire-land happened to be travelling close by the shore of Loch Ness, when he heard the fishermen there tell of a strange beast. This crea-ture was said to devour fishermen from time to time. St. Columba therefore said prayers and succeeded in driving away that monster.

In 1572, the Loch Ness monster appeared once more. It was described as having a pointed tail resembling a sickle, and was reported to have swallowed three fishermen. In those days, the monster seemed to be of interest only to the people who lived by Loch Ness and fished in its waters. In 1933, however, after a road around the shoreline of the lake had been opened, the Loch Ness monster became a popular topic all over Britain, because laborers who had been brought in to build the road took tales of the monster back to their home towns and villages.

Then, in 1934, a London surgeon, on a sightseeing trip in Scotland, took a snapshot of the purported monster. The creature appeared to have a long, slender neck projecting above the surface of the water. By the end of the 1950's, reports of the Loch Ness monster had even begun appearing in Japanese newspapers. From time to time thereafter, people reported seeing "Nessie," but no particular excitement was aroused by these accounts. The Loch Ness monster was treated as one of the "sensations of the century" only after America had come down with the anti-science disease and had begun to spread it all over the free world. The beginning of the "Nessie" craze was an episode that occurred on the evening of August 21, 1966. The crew members of a ferryboat, the *Farma*, that was proceeding along the lake near Urquhart Castle, saw a creature with what appeared to be three humps. When they looked at the vessel's radar, it showed five humps projecting above the surface of the water, and moving alongside the ferryboat, in the same direction, for about 30 minutes. The vessel's captain, George Ralph, later said:

> I have gone back and forth on this run hundreds of times since 1945, but that was the first time I ever saw anything so weird. I can't say definitely that I saw a monster, but it was certainly un-canny, and it's a mystery as far as I'm concerned.

Two years later, in 1968, scientists of the University of Birmingham (England) formed an investigative team to go to Loch Ness. In 1969, Dan Taylor, an American, arrived with a midget submarine. This 28-year-old explorer made a number of dives during the summer of that year, but all he saw were small rainbow trout. In the spring of 1971, the manufacturers of Cutty Sark whiskey, in cooperation with Lloyd's of London, decided to offer a prize of one million pounds (equivalent at the time to about $2.5 million) to anyone who could capture "Nessie" alive during the one-year period from May 1, 1971, to April 30, 1972. Just when the time limit for that reward had nearly expired, a remarkable occurrence took place (March 31, 1972): the dead body of a creature that appeared to be an infant "Nessie" was discovered at Loch Ness. Scientists who happened to be there conducting a study excitedly proclaimed it the actual remains of an aquatic monster. But it was revealed the next day to be an April Fool's Day prank. Someone had taken the corpse of a walrus that had died in a nearby zoo, and changed its appearance by shaving off its whiskers, filing its teeth to points, and stuffing its cheeks with gravel. It was then dumped on the shore of Loch Ness.

In 1973, a fifteen-person investigative team arrived from Japan, led by Shintaro Ishihara, a well-known novelist and conservative member of Parliament. The group studied the depths of the lake from September to November of that year, but found nothing. In 1975, an American investigative group was sent to Loch Ness by the Academy of Applied Sciences, a privately-funded organization in Boston. This team, utilizing sonar and underwater cameras, claimed to have succeeded in taking an underwater photograph of what appeared to be a strange creature. The photograph in question was to be exhibited at what was planned as a two-day scientific symposium to open in Edinburgh on December 9, 1975. Advance reports in the press and on television created such a stir, however, that the symposium was cancelled, and the photograph was released on December 10 by Dr. Robert Rines, of the Academy of Applied Sciences. The picture proved to be extremely ambiguous, and it was impossible to tell whether it actually showed a Loch Ness monster.

According to Rines, the photograph had been taken on June 20 at a depth of about 40 feet. The expedition had continuously scanned the waters of the lake with sonar, and whenever anything appeared to be moving, pictures were taken using strobo-scopic ("strobe") lights. From the photograph, Rines estimated that the creature had a 12-foot neck, and a body about 65 feet in length. This photograph was reproduced in newspapers and magazines all over the world. It appeared in the December 11 issue of the British science magazine *Nature*, which even awarded "Nessie" the scientific name of *Nessiteras rhombopteryx*: "Nessi-" for the lake, "-teras" meaning "frightful" in Greek, and "rhombopteryx" meaning "diamond-shaped fin" in Greek. The latter word referred to what looked like a four-sided fin or flipper in Rines's photograph. However, two theories refuting the apparent photograph of "Nessie" immediately emerged. One theory was that it was a photograph of a model that had been used in shooting a motion picture. On December 12, 1975, the day after the release of the photograph, it was pointed out by Roy Muir, a retired Scottish librarian, that in the late 1960's a film had been made on the shore of Loch Ness. A model of the Loch Ness monster, used in that film, had sunk to the bottom of the lake, and might well be the object seen in the photograph.

In 1969, an American company, Mirisch Productions, had filmed *The Adventures of Sherlock Holmes* at Loch Ness. In one scene, the famous detective and his assistant Dr. Watson, in the course of solving a case, rowed at night across Loch Ness. Suddenly the monster appeared, overturning their boat. The "monster" in this scene was quite large; it was made of plastic over a wire frame, was tied to a midget submarine about 50 feet long, and had a 10-foot neck. On July 21, 1969, however, this particular incarnation of "Nessie" had broken its towline, overturned, and plunged to the bottom of Loch Ness. Muir contended that what the Rines expedition had photographed was the Hollywood version of the creature, and that the head and other portions visible in the picture looked just like the plastic model.

The other counter-theory about Rines's photograph was that it merely showed a dead cow. Joseph Harwood, the president of the British Underwater Photography Association, suggested that the submerged remains of a cow that had drowned in Loch Ness had been mistaken for the monster. Robert Rines, however, did not accept either of these arguments. The following year, in 1976, he again visited Loch Ness. This time, America's most prestigious newspaper, the *New York Times*, was backing the expedition. Despite the use of sonar, infra-red detectors, and other advanced equipment, however, no trace of the mystery creature was found.

The photograph taken by Rines in 1975 was basically too blurred to justify the view that it depicted the Loch Ness creature. Moreover, although the picture taken by the English surgeon in 1934 appears to show a long neck protruding above the surface of the water, there is not a single other object included in the picture with which the length of the neck can be compared, nor does anything in the photograph actually identify the location as Loch Ness. Some scientists have even said it shows the tail fin of a porpoise breaking the surface of the ocean. By their very nature, photographs can be made by a variety of techniques and artifices, and countless trick photographs have been taken. Photography is a great help when doing scientific investigations and research, but it is not very scientific to proclaim the existence of a living prehistoric creature on the basis of a single blurred photograph. If there actually were a creature of such great bodily size in Loch Ness, surely someone would have discovered the remains of salmon it had eaten, or even found droppings of some sort. If it really were some species of long-necked reptile that had survived from remote prehistoric times, then hundreds or thousands of generations of the creature would already have been born and died so that the bottom of Loch Ness would have to be piled with layer upon layer of the giant skeletons of the current "Nessie's" ancestors. But not one bone or fossil of any kind has yet been discovered there.

From the standpoint of the science of biology, it is quite inconceivable that a so-called monster should be living in Loch Ness; but as the anti-science disease has grown more virulent, people have become less able to look at things from a scientific point of view. It is most unfortunate, in my view, that a respected newspaper and a science magazine with over a century of tradition can become caught up in this kind of furor, and play a role in encouraging mass delusions among the public.

Psychic Magic

Even more interesting than the excitement about UFO's and the Loch Ness monster was the stir created by the so-called "paranormal" boom, created by Uri Geller, which swept over the countries of the free world from the autumn of 1973 through 1974. The point of origin was America. Using his "psychic power," Uri Geller was apparently able to bend metal spoons, start watches that had broken down and stopped running, and describe pictures inside sealed envelopes. Geller was born on December 20, 1946, in Tel Aviv, Israel. In his autobiography, Geller recalls a mystical experience he had at the age of three or four when he was playing alone in a garden near his home in Tel Aviv. Suddenly there was a high-pitched ringing in his ears. All other sounds ceased, and the trees stopped moving in the wind, "as if time stood still." Looking up at the sky, he saw "a silvery mass of light" coming toward him. He felt as if he had been knocked over backward, there was a sharp pain in his forehead, and he lost consciousness completely.

Geller claimed to have gained his "paranormal" abilities as a result of that mystical experience, and said that around his eighth or ninth year he became able to bend spoons and silver dishes merely by using his willpower. After he completed his university studies, he worked for a while on a farm and served in the armed forces. In March, 1970, he became an entertainer, and began showing audiences his spoon-bending ability. One theater manager soon told him "there ought to be more to the act," and began helping Geller with a stage trick. The manager would watch people get out of their cars before they entered the theater, and would write down their license numbers. He would then have them ushered to selected seats as they came into the theater. Before the show, Geller was given a list of license numbers and corresponding seat locations, and during the "demonstration" of his psychic powers, he would point to those people and tell them their license plate numbers. In his autobiography, Geller said that he hated himself every time he did the license plate trick, but his performances quickly made him famous in Israel.

An American physician and neurophysiologist, Dr. Andrija Puharich, heard about Geller and became interested in his feats, and in August, 1971, went to Israel. He became convinced that Geller's powers were those of an envoy from an advanced civilization or a world outside our own, and brought the young Israeli to America in

1972. The first thing Geller did there was to submit to a series of "scientific" tests administered at the Stanford Research Institute in California.

The results of those tests appeared to indicate that Geller was indeed gifted with "paranormal" abilities. Dice were sealed inside a metal box, and thrown by shaking the box. Geller was then asked to tell what the dice would show. Out of ten throws of the dice, he said twice that he did not know, but named the value of the dice correctly the other eight times. He also was successful in describing pictures inside sealed envelopes. On September 25, 1973, Geller gave his first public performance in America, at Town Hall in New York City. His demonstration mainly involved starting stopped watches and bending rings, but according to the next day's newspapers, he scored a tremendous success with his audience. What followed amounted to a "Geller whirlwind." Invitations for appearances came rolling in, from such countries as France, Britain, West Germany, Sweden, and Denmark. Whenever Geller performed on television, viewers would phone the studio to report that spoons in their own homes had been bent. This storm of enthusiasm swept over Japan as well. In December, 1973, Nippon TV, a Japanese commercial television network, videotaped a Geller demonstration of "paranormal" abilities, and later invited him to visit Japan, in February, 1974. He demonstrated spoon-bending, and also broke the handles off spoons, using his strange powers.

A number of Japanese children, who watched Geller on television, became fascinated by his feats and began imitating them. One such child, an 11-year old boy, was given great publicity on television and in popular weekly magazines. His father declared in print that no tricks or deceptions were involved. Even some scientists joined in the "paranormal" boom. One professor of engineering at a national research institute said, "I tried to do it myself, and found that I was able to bend a fork. To say that this is not a fact would go against my conscience as a scientist." He reported in all seriousness that he had checked the bent fork and found it 0.003 grams lighter than before, and that the point of bending had become magnetized.

But despite the widespread public acceptance of Geller's feats, the *Asahi Shimbun's* daily front-page column, "Vox Populi, Vox Dei," raised objections:

> A magician glibly assures us that he has nothing up his sleeve, and then proceeds to amaze us by pulling doves, goldfish bowls, and rabbits out of a silk hat. But no one really believes the magician has "paranormal" powers. We all realize that he in fact does have something up his sleeve, that he is using trickery and deception. But because we cannot figure out how he does it, the magician is able to make a living entertaining us. Those "paranormal" feats that the public cannot figure out are just like the magician's sleight-of-hand, but people take them seriously, using words like "miraculous" or "mystical." A trick is a trick, after all, and we cannot let ourselves fall victim to mass hypnosis.

This was written by Junro Fukashiro of the paper's editorial staff, and appeared in the *Asahi Shimbun* on April 20, 1974. As expected, the column evoked a variety of responses from large numbers of readers. Most of them were along these lines: "People think that anything they cannot understand must be some kind of trick. That is because they have a blind faith in science. But the world is full of things that science cannot explain." One of these letters said: "My son can make spoons bend. Please visit us and see for yourself. Once you see it, you too will be able to believe in paranormal powers." Fukashiro decided ᵗo visit the child's home, in the suburbs of Tokyo. At the time, I was deputy chief of the Asahi's science department, and Fukashiro asked me to accompany him. The child's mother told us: "Really, it gives me an uncanny feeling. . . . When he is concentrating, he says it's as if fireworks are exploding all around his head. It's like the air is full of electricity, and it jumps from his fingertips to the spoons, and the power of that electricity makes them bend."

We had brought five spoons with us. Two of them were solidly made, too heavy for even an adult to bend. The boy began by setting aside those two spoons. "My mental power won't work on those two spoons," he informed us. Then he led us out into the garden behind the house, and explained: "I'll keep my left hand raised in the air, and bend the spoon just by holding it in my right hand. I'll bend it without using my hand." He raised his left hand, and squatted down with his back to us. He appeared to be concentrating mentally, and after a couple of minutes he suddenly shouted, "Bend!" He then tossed the spoon aside. The spoon had indeed been bent.

Fukashiro asked the boy, "How about doing it facing us, instead of with your back to us?" "That's no good. If other people are watching, I can't concentrate," replied the boy. We had realized, from watching his movements, that he was bending the spoon by pressing it against his abdomen with his right hand. The bowl of the spoon was round, so that it was not all that painful when forced against the body. Using the handle of the spoon like a lever, by holding it in his hand, he was able to exert great force upon the narrowest portion of the handle and bend it easily. At around the same time, and independently of Junro Fukashiro's investigation, the editors of *Shukan Asahi* (a weekly magazine published by the Asahi newspaper company) asked to take documentary photographs of spoon-bending performed by the best-known of the "paranormal" children. So that the photographs would be clear, the spoons were painted white. When the test was over and the boy stood up, numerous imprints of white-painted spoons were found in the black cloth that had been spread on the floor. The room had been darkened in order to shoot the pictures, and the boy had taken advantage of this to press the spoons against the floor in order to bend them.

In the May 24, 1975, issue of *Shukan Asahi*, spoon-bending was finally exposed by scientific tests, in an article entitled "The Final Blow to the 'Paranormal' Boom." The Nippon TV network, joined by several other weekly magazines, counterattacked along these lines: The "paranormal" children may, out of fatigue, have resorted once or twice to deception. But it would be unscientific to deny all of their feats just because of one instance of trickery. And even if we assume that the children are fraudulent, Uri Geller is certainly genuine. It appeared, however, that the majority of the public agreed with the article in the *Shukan Asahi*, and the "paranormal boom" in Japan gradually faded away.

In Europe, on the other hand, the craze continued. The October 18, 1974, issue of the British science magazine *Nature* carried a report by the Stanford Research Institute. But, another British science journal, the *New Scientist*, (October 18, 1974) published an essay by the physicist, Dr. Joseph Hanlon, in which he declared that Uri Geller's "paranormal" powers were nothing but a patent on a super-miniaturized radio transmitter-receiver. This device is small enough to be concealed inside a hollow tooth, Hanlon declared, so that Geller could have used such a radio to communicate with an accomplice. The article by Hanlon is considered to have been a major influence in dampening the enthusiasm for Geller's claims.

This "paranormal" silliness was touched off in the United States. The "paranormal" craze could flourish only because the public had caught the anti-science disease, and had lost the ability to look at the world and its phenomena in an objective, scientific manner.

Old-Time Religion

The gravity of this anti-scientific problem was indicated by a court judgement rendered in California in March, 1981. The plaintiff, Kelley Segraves, a so-called "creationist," considered it wrong for schools to teach only the Darwinian theory of evolution, and believed that the Biblical theory of creation should also be taught in biology classes. Of course, the debate goes back into American religious history, but it had a new twist.

According to "scientific knowledge," a cosmic explosion known as the "big bang" occurred between 15 and 20 billion years ago, marking the origin of the universe in which we live. The sun and the earth were formed about 4.5 or 5 billion years ago, while the earliest forms of life first appeared in the earth's oceans about 4 billion years ago. Life gradually evolved through fish, amphibians, and reptiles, until several million years ago, when the earliest ancestors of mankind began to appear. This is the accepted scientific description of the origin of the universe and of life, and it is this evolutionary account that high school science classes have traditionally been taught.

Like Kelley Segraves, however, many people do not accept this style of explanation of the origins of man and the universe, for it is contrary to their beliefs as Fundamentalist Christians. Fundamentalism, which became a major force during the decade after the first World War, teaches the literal truth of the Bible, and accepts the Biblical version of the Creation, as well as the various Old and New Testament accounts of miracles. According to the Biblical chronology followed by the Fundamentalists, the world began only about 6,000 years ago, when the earth, the sun, the stars, and all the plants and animals were created by God in the period of five days. Man was created on the sixth day. Fundamentalists express their position as: "Man was created, by the grace of God, as a human being from the beginning, and did not evolve from some monkeylike creature. Darwin's theory of evolution is nothing more than a hypothesis, and yet it is being taught in our schools as if it were an absolute truth. Children are completely bewildered at being taught something totally different from what they have been told in their churches and in their own homes. The Biblical story of Creation should be taught in biology classes, on an equal basis with other theories of evolution."

On March 3, Kelley Segraves' son Casey, a 13-year-old junior high school student, testified before the court as a witness. "I don't believe man is descended from apes, or fish, or reptiles," he said. "I believe that God created man as a man and put him on the earth." He went on to describe what he had been taught in school: "The teacher taught us that man was descended from apes. I didn't believe that, but on tests, I had to answer that man was descended from apes."

This was not the first American courtroom drama involving evolution and its opponents. The first and most famous evolution trial took place in Dayton, Tennessee, in 1925, the famous Scopes trial. In March of that year, the Tennessee state legislature had passed a law making it illegal to teach any theory of creation other than the account found in the Old Testament. However, a high school teacher named John T. Scopes had broken the new law by teaching Darwin's theory of evolution to his biology classes. The question at issue when Scopes came to trial was not whether the Darwinian theory was scientifically valid, or whether Tennessee's anti-evolution law was constitutional. Rather, the only issue was whether Scopes had actually taught evolution in a public school classroom. Naturally, Scopes admitted that he had indeed done so, and the court duly fined him $100. Scopes immediately appealed the verdict to the state Supreme Court. The latter body upheld the constitutionality of the 1925 law, but acquitted Scopes on a technicality. The Scopes "monkey trial" received tremendous publicity at the time, and is best-known for its courtroom debates between the famous attorney Clarence Darrow, for the defense, and the noted political orator and Fundamentalist William Jennings Bryan, as the star prosecution witness. There were no subsequent trials of this kind, and attacks on evolutionary theory appeared to have become a thing of the past.

But around 1969, when anti-science attitudes became strong again, a movement began in the state of California to alter the curriculum of high school biology classes. By 1972, the press and television were giving this movement considerable attention, and the publicity certainly increased the influence of the Fundamentalists. During the 1980 Presidential election compaign, Ronald Reagan declared, "There are great flaws in the theory of evolution," thus appearing to take a position sympathetic to the Fundamentalists. In the 1981 evolution trial in California, Kelley Segraves avoided an all-out attack on the theory of evolution itself. Instead, he simply sought to have the words "most scientists believe" added to the description of evolutionary theory in the California state curriculum, and so the matter was quickly disposed of. The judge's verdict stated that California's policy on teaching evolution did not violate the religious rights of Fundamentalists. The state board of education, however, was ordered by the judge to transmit to all high schools an amendment to its long-standing policy statement on the teaching of evolution. The revised policy was to state that evolution should not be taught as dogmatic, irrefutable fact, but rather as a scientific theory. This verdict did not deal any major blows to the theory of evolution. But because of all the media publicity, an unexpected chain reaction occurred. Legislation incorporating the creationist viewpoint was submitted in at least fifteen state legislatures.

On March 18, 1981, the Arkansas state house of representatives, by an overwhelming majority of 69 to 18, approved legislation requiring that public schools teach, along with the theory of evolution, the theory that God created man. This meant not merely discussing both theories in history or social studies classes, but giving equal time to each theory in biology courses as well. This did not happen back in the nineteenth century, but in the year 1981. To the average Japanese, such a law seems like sheer madness. Immediately after the Arkansas legislature enacted this statute, the American Civil Liberties Union filed suit seeking to have the law declared unconstitutional. The Arkansas law, however, was immediately thrown out by the State Supreme Court in January 1983 for lack of educational merit.

The well-known American science writer Isaac Asimov, greatly distressed at this situation, wrote a long article for the *New York Times Magazine* in June, 1981. The following are some excerpts:

Do you suppose [the creationists'] devotion to fairness is such that they will give equal time to evolution in their churches?

Are we now . . . to ride backward into the past under the same tattered banner of orthodoxy? With creationism in the saddle, American science will wither. We will raise a generation of ignoramuses ill-equipped to run the industry of tomorrow, much less to generate the new advances of the days after tomorrow.

We will inevitably recede into the backwater of civilization and those nations that retain open scientific thought will take over the leadership of the world and the cutting edge of human advancement.

When trust in scientific thought goes into a decline, people turn to religious belief. Perhaps for that reason, America experienced a boom in sales of religious-type publications, beginning around 1970. On May 12, 1981, the *New York Times* carried an article headlined "Religious Publishing: Going Skyward." It quoted the vice-president of one religious publishing house as saying of this boom, "In the last five or ten years it's gotten much larger than even people in the trade thought it would get." Until about ten years ago, no author of religious books could earn a living from royalties alone. Today, however, many such writers are able to support themselves very well with their books. Billy Graham's book *How To Be Born Again* was first published in 1977, with an initial printing of 800,000 copies, and it has already sold more than one million copies. Graham's most recent volume, *Till Armageddon*, also became a best-seller. A religious book entitled *The Late Great Planet Earth*, by Hal Lindsey, is said to have sold 15 million copies. Other best-selling books discuss subjects like marriage, childrearing, and other human relationships, from a religious point of view. Whatever the causes, this revival in religion is linked to a number of other factors during this period that were not supportive of the trust in science and technology that were needed to keep America strong and economically competitive.

VII
WASHINGTON'S SCIENCE BLUNDERS

The tragedy of American science policy, but an intrinsic part of the democratic policy process, has been the political manipulation of science and technology budgets. In a democratic nation, all elected officials from presidents to minor municipal politicians must submit to judgement at the ballot box. Once defeated at the polls, even the most powerful legislator becomes just a common private citizen. Politicians naturally tend to be very sensitive to public opinion trends. If an "environmental boom," for example, sweeps the country, politicians discover that making statements on such issues will get them into the media, increasing their chances of success at the polls. In the early 1970's, many politicians, without taking the trouble to check the scientific evidence, spoke out indiscriminately in support of "environmental" causes. As intellectuals and the media grew more hostile toward science and technology, politicians naturally came to share that antipathy. During the period when the anti-science disease was growing increasingly severe, it became almost impossible to find anyone in non-scientific, educated circles who would speak up in defense of science and technology.

This climate of anti-science influenced even the President of the United States. Edward J. Burger, Jr., who served for six years in the Office of Science and Technology at the White House, wrote the following in his book, *Science in the White House*:

> During those times when a nation warmly embraces science and tech-
> nology and sees them as desirable partners in the enhancement of the
> economy or the advancement of the physical frontiers, a President finds
> it to his advantage to take active steps to look to the scientific commu-
> nity for help in formulating national decisions. A popular view of sci-
> ence and technology as antithetical to national well-being, however,
> discourages any manifestly close relationship between the White House
> and science.

On January 26, 1973, President Nixon sent Congress a message on restructuring the Executive Office of the President. He eliminated the White House post of Science Adviser, and abolished the Office of Science and Technology. In addition, the members of the President's Science Advisory Committee were required to submit their resignations.

The post of Presidential Science Adviser had been created by President Eisenhower in October 1957, at a time when the USSR had just launched the first man-made satellite, *Sputnik 1*, and the technology gap between the Soviets and the United States had become a major issue. The Science Adviser was able to meet with the President at any time, to convey directly his views on both military and peaceful aspects of science and technology. From the creation of this post until it was abolished in 1973, David Z. Beckler served as principal assistant to all six Presidential Science Advisers. He wrote:

> Thus, in one fell swoop, the President eliminated the entire White House science and technology mechanism that had been painstakingly erected in the wars following the Soviet Sputnik in 1957. Unfortunately, the President's action did not reflect a careful assessment of the strengths and weaknesses, past accomplishments and future potential of the science and technology mechanism in the White House. Rather, it appeared to be the result of a hasty decision taken on the basis of general considerations.

Burger and Beckler offer their own interesting analyses of why President Nixon expelled the scientists from the White House. According to Burger, the President's firing of his Science Adviser was the ultimate consequence of the fact that the politicians and the scientists differed fundamentally in their respective ways of thinking about issues. The gap between the two groups widened rapidly with the rise of the consumer movement and the anti-pollution movement, resulting finally in conflicts that ended in a defeat for the scientists.

Political decisions are not scientific in nature, being generally a product of compromise and judgement. The competing demands of different interest groups, each desiring benefits for itself, must be astutely balanced by seeking areas of agreement in order to formulate policy. That is the traditional *modus operandi* of politics, which does not require that a problem be analyzed scientifically and the technologically most appropriate solution be adopted. Politicians must accommodate the demands of the greatest possible number of voters, and are thus often driven to select unscientific alternatives. The more serious the problem, moreover, the less time that may be available to reach a decision. In many cases, the policy ultimately adopted is intended primarily to placate the public's fears or anger, or to divert the attention of the electorate away from the real and less soluble problem.

Someone trained as a scientist or an engineer, however, is likely to be totally unaccustomed to evasive and opportunistic compromise solutions to problems, for scientific research involves the exacting and rigorous pursuit of truth. Technology is equally demanding and precise. A minor flaw in design or construction can result in a breakdown or even a disaster. For that reason, scientists and engineers constantly strive to eliminate subjectivity from their thinking, and to be as objective as possible when analyzing problems. Whenever factors can be quantified, it is done in order to make possible an objective comparison. Thus, the views that a scientific adviser brings to the President have the characteristics of scientific analysis. The adviser ideally picks out various policy options among which the President can choose, then quantifies and compares them, analyzes their likely effects, and presents them to the President. That, essentially, is the task of the science adviser.

Sometimes, however, the results of this kind of objective analysis collide with the needs of interest groups or public policy. Moreover, objective analysis takes time, so that painstaking analysis may not be suitable in a crisis. In some cases, scientific analysis can even interfere with decisionmaking.

One example cited by Burger is the study conducted for the proposed "New Technology Opportunity Program." For the 1972 election campaign, President Nixon wanted to announce an official initiative in an area of technological development that would directly benefit the electorate. A survey for this purpose was begun in July 1971, with Science Adviser Edward David in charge. He solicited ideas from every government agency, classified them, and winnowed them down to fifty ideas that became the first-stage candidates for selection. Those fifty were submitted to the President in an interim report in December 1971. This was followed, however, by a further process of even more detailed scientific and technical scrutiny, which determined that not one of the fifty ideas appeared to be attractive as a possible "Presidential initiative" in technology. President Nixon must have received the impression that whenever scientists study issues with scientific care, they leave nothing politically attractive remaining at the end.

The plan to prepare an "Annual Report on Science and Technology" met a similar fate. The idea was to publish an annual report that would demonstrate the role of science and technology in bringing benefits to the public. Work on the project began in late 1970, and after about a year, a report of around 450 pages had been compiled. In the end, however, the report was never published. Because of the excessively scientific judiciousness and deliberation with which the document had been prepared, it touched on several politically controversial topics, and protests were voiced that unpleasant political consequences would ensue if the report were ever issued. This episode, too, must have made President Nixon feel that scientists should not be entrusted with such tasks.

This impatience or irritation with scientists was not confined solely to the President. The scientists, for their part, were angered by a number of things. For instance, there was the Muskie amendment to the Clean Air Act of 1969. The White House Office of Science and Technology opposed as "unscientific" the amendment's restrictions on automobile exhaust emissions. The relationship between automobile exhaust gases and air pollution was not yet well understood from a scientific standpoint. Thus, the emission control standards had been determined on the basis of the unscientific assumption that air pollution was directly proportional to the quantity of exhaust gases emitted. Moreover, from a technological standpoint, no one had any idea whether automobiles could even be built to comply with such emission standards. If the standards were rushed into effect the cost would be extremely high. There was absolutely no scientific evidence as to whether air quality would actually improve enough to justify the costs of the emission standards.

The Office of Science and Technology made known these views. But in the "environmental protection boom" prevailing at the time, and given the political ambitions of members of Congress who wanted to take up the environmental cause, the scientists' arguments were inevitably dismissed as mere nitpicking. The President's Science Adviser deplored the fact that in the heat of political enthusiasm, a succession of laws related to environmental issues was enacted, none of which had been subjected to scientific scrutiny. The politicians, on the other hand, were convinced that timing, in the political sense, would be lost if cautious and time-consuming scientific studies were allowed to continue. In this way, the gulf between the scientists and the politicians continued to deepen, Burger believes, until the scientists were finally edged from the White House.

Beckler, by contrast, suggests factors that are perhaps more prosaic. First, there was the opposition expressed by many intellectuals, including scientists, toward the war in Vietnam. During that period, the Science Adviser and the scientists in the Office of Science and Technology (OST) were viewed by many administration officials as enemies who had burrowed into the White House. The second reason mentioned by Beckler is that the chairman of the Supersonic Transport (SST) Panel of the President's Science Advisory Committee openly opposed development of the SST. President Nixon wanted to push forward to develop the SST, but he was confronted by this opponent who emerged under the White House roof. Third, according to Beckler, one scientist who had served on the President's Science Advisory Committee had sided with those opposing deployment of an anti-ballistic missile (ABM). This, too, was contrary to President Nixon's own views. Thus, one provocation followed another, until President Nixon finally banished scientists from the White House.

The Space Program Stalls

Nearly all of America's large-scale technology suffered from the effects of these anti-scientific sentiments, but it was the space program that perhaps endured the severest blows. We have already seen in Chapter IV that when the first two astronauts reached the moon's surface in *Apollo 11*'s lunar landing module, some American opinion-molders expressed a strong sense of unease. For the United States, *Apollo 11* marked a turning point. NASA had been able to push forward energetically with the manned lunar landing program because there was an overriding goal, on which the nation's prestige depended, of catching up with and passing the USSR in the race for space. America passed the USSR, but discovered also that the Soviets seemed to drop out of the race.

The USSR's annual gross national product was only about half that of the United States at the time, and the Soviets apparently could not afford the huge expenditures demanded by a manned lunar landing program. At an early stage, the Soviet Union abandoned any thought of a manned landing, and concentrated on using automatic probes to bring back samples of lunar soil and rock. Three days before the launching of *Apollo 11*, the Soviet Union launched *Luna 15*, an unmanned lunar probe, designed only to return from the moon with soil and rock samples. *Luna 15* failed to achieve a "soft" landing on the moon's surface, but the next such vehicle, *Luna 16*, took off on September 12, 1970, and successfully returned to earth with the desired samples.

This came as a major surprise to Washington. The United States had thought the Soviets had been soundly beaten, but now they discovered that the Soviets were competitive in a different race. The United States had been engaged in a contest with a non-contesting rival. True, America had been declared the winner, but it was in a sense a victory by default, and the aftermath was somehow a sense of hollow anticlimax. NASA had planned a number of very impressive projects to succeed the Apollo program. Some examples: erecting on the surface of the moon a semipermanent base where people could live; sending a manned landing vehicle to Mars; and constructing a giant space station in orbit around the earth.

In December, 1965, I visited the Marshall Space Flight Center in Huntsville, Alabama. The director at the time was the German-born scientist and engineer, Wernher von Braun, who played a major role in developing the V-2 rocket for Nazi Germany during the second World War, and then had come to America. I interviewed Braun in his office at the space center. At the time, the two-man Gemini spacecraft had made history's first successful orbital rendezvous, but the Apollo spacecraft had not yet been completed. Braun, however, was already thinking beyond the Apollo project. He replied to my questions in these words:

> The next stage after the moon landing project, I believe, will be to construct a large space station in orbit around the earth. A large number of scientists and technicians would live there, and make oceanographic and weather reports, do astronomical and biological research, and so forth. I think the best size would be a station where a dozen or two dozen or so people could live. After that, we will begin to explore Mars and Venus. We would use a Saturn V rocket to lift the components into earth orbit, and by rendezvous and docking we would assemble them into a large spaceship, which would then go to Mars and Venus.
>
> Question: How long would it be before this becomes a reality?
>
> In 1975, a three-man spacecraft could pass close to Venus, and in 1978 another three-man craft could pass close to Mars and return to earth. Then, in 1982, an eight-man spaceship could fly to Mars, and four out of the eight could actually land on Mars and explore the surface for 20 days. All of this can be done by using the rockets and spacecraft developed for the Apollo project, along with our launching pads and tracking facilities, plus the nuclear-powered rocket now being developed.
>
> Question: But might this not be too visionary?
>
> This hasn't been adopted as a formal program by NASA, but money is already being spent for research related to these plans, and the research is fairly far along. For instance, if we decide to land human beings on Mars, within 90 days we can launch ten Saturn V rockets and assemble in earth orbit a spaceship to use in traveling to Mars. If the spaceship leaves earth orbit on December 28, 1981, it will land on Mars on August 4, 1982, leave Mars 20 days later, and return to earth on March 29, 1983. That's a grand total of 456 days for the journey.

Von Braun's dreams had already soared beyond the moon to Venus and Mars. I was greatly impressed by his enthusiasm — he even specified the date 15 years in the future on which the long journey to Mars would begin.

Apollo 11 lifted off from Kennedy Space Center at Cape Canaveral on July 16, 1969. On that day, Vice President Spiro Agnew said that after the moon, America's next target would be Mars. The same day, in a press conference held at Kennedy Space Center, Wernher von Braun said this:

The *Saturn V* has one capability with respect to the Moon that is not mentioned very frequently. If we decide to have extended, even permanent activities on the lunar surface, it would be highly attractive to use the *Saturn V*, and also on one-way unmanned missions to fly more cargo and consumables to the lunar surface. In this fashion, we could soft-land approximately 38,000 pounds of net payload on each *Saturn V* flight, and we could probably further increase that load to close to 50,000 pounds by a few rather nominal improvements. Now, in addition to this, I think the *Saturn V* can also be used very effectively as a logistics supply vehicle for planetary flights, if what the Vice President indicated today should come to pass and this country one day commits itself to a manned landing on Mars. We then would probably use *Saturn V*'s to carry parts of interplanetary space vehicles and propellants for these vehicles into a low orbit around the Earth, where the interplanetary spaceship would be assembled. This interplanetary vehicle will probably be powered with a nuclear engine, but we have no plans ever to take off with a nuclear engine from this facility. We would prefer to start up a nuclear engine only from orbit, and so the *Saturn V* would be a logistics carrier to bring all the weight — the crews, the propellants, the expedition equipment — into earth orbit, where it would be assembled.

Von Braun became Deputy Administrator of NASA in February 1970, and was transferred to NASA's Washington headquarters. He tried to persuade Congress to support the project for a manned landing on Mars. He exerted his utmost efforts in that cause. The dream of sending human beings to Mars was one he had cherished since boyhood. He wanted somehow to make the Mars voyage a national commitment of the United States. But by the time he began trying to convince Congress, America had already come down with doubts about science. The public had begun to insist that if there was money to spend on space exploration, it would be better spent on preventing pollution and ending poverty here on earth. America turned inward.

For a scientist like Braun, attempting to sell an expensive space program to a skeptical Congress was certainly not a congenial task. But he set about trying to persuade the members of Congress one by one, in terms like these:

All of the money spent on the moon landings was spent right here on earth. Our efforts in space development have produced all kinds of inventions and technological innovations, and America can now take pride in having the highest level of science and technology in the world. If we were to dismantle the superb organization already created by NASA, it would be a major loss to this country. We urgently need a commitment to the program for the Mars voyage.

But the Senators and Congressmen remained unmoved. Instead, American society as a whole had already begun to grow hostile to science and technology. The grand plan to send mankind on a voyage to Mars was not approved by Congress. Instead, Congress voted only to support the space shuttle, the reusable winged space vehicle that now goes into orbit around the earth.

There are various opinions on the relationship between the space program and technological progress. For example, Masaru Ibuka, the founder of the Sony Corporation in Japan, said the following in an August, 1981 interview with a reporter for Tokyo's *Asahi Shimbun*:

> There was something I said perhaps 20 years ago, when America's power and prestige were still bright. I tossed a stone at them, as a warning, by saying that America's electronics industry would become spoiled, eventually, because of defense and space. I'm sure everyone just thought: there's that fellow Ibuka talking nonsense again. The only real reaction was a cheer from one American scholar.
> What led me to make that observation at the time? It was because America's electronics industry was tending to become dependent on defense and space contracts. If the fate of the country had been at stake, it would have been different — but otherwise, engineers being only human, they tended to take the easy way out. They had an attitude of "let Uncle Sam foot the bill." I thought this easy-going attitude might weaken the electronics industry in America. Looking back on it now, I think you'll have to admit that my warning wasn't so far off target.

In retrospect Ibuka's views are certainly worth heeding. Money spent by the national treasury without competition, inevitably tends to be used carelessly. When private moneymaking corporations spend their own hard-earned profits on research, the researchers are always conscious that not one dollar should be wasted. Figure 15 shows what percentages of research and development expenditures in 1967 and 1975 in five countries came from the private sector and from government. In the case of the

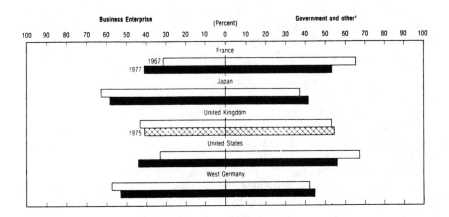

Figure 15   PERCENT DISTRIBUTION OF GROSS EXPENDITURES ON RESEARCH AND DEVELOPMENT BY SOURCE OF FUNDS

United States, government expenditures constitute the major source of research funds, by a ratio of about 60 to 40. Japan is just the reverse, with a 60 to 40 ratio in favor of private enterprise funding.

It is probably not inaccurate to say that the diligence of researchers and the development of commercial products is influenced to some extent by whether their funds come out of their own or the taxpayers' pockets. Military research and development, moreover, are inevitably accompanied by a need to preserve secrecy, with the consequence that inventions and innovations resulting from military research do not flow easily into the private sector. It is only years later when security restrictions are lifted that American companies are allowed to use this new technology. By then, other countries are likely to have succeeded in developing similar innovations. In this American firms lose their head start and their ability to compete in the world market.

In the case of the space program, however, the area covered by security restrictions was much more limited, and a variety of inventions and innovations quickly flowed into the private sector. For example, freeze-drying techniques were developed in order to produce lightweight, nonperishable food for the astronauts in space. This process was immediately applied to the manufacture of instant coffee, instant orange juice, and similar products. Integrated circuit technology was developed in order to miniaturize spacecraft guidance and control systems, and almost immediately came into use in pocket calculators, auto-focusing systems for cameras, and countless other applications. In taking technological innovations like these and applying them to consumer goods for the mass market, American firms have been far surpassed and outdistanced by their Japanese rivals; but at the same time, America has been able to reap large patent and licensing royalties from Japan. The American space program also gave birth to a large number of inventions, innovations, and technological improvements. The space program, in a sense, served as a pipeline through which huge subsidies flowed to the private sector.

In 1965, I visited a number of NASA research institutes and space centers, and wherever I went, I saw large rooms packed with rows and rows of high-speed computers. I sighed and thought to myself that American computer manufactures are going to keep growing steadily. It looks as if Japan's computer industry doesn't have any chance of catching up with America. But I was wrong. The defeat of America's electronics industry by Japanese companies was not caused by American over-spending on the space program. Rather, it was a result of the 1970 decision to allow the space program to die. America's private industry had acquired the habit of depending on government projects, so that when "finis" was written on the enormous space program, American industry was like an aircraft whose fuel had suddenly been cut off: it lost speed and stalled out.

If President Nixon and the Congress had followed the advice of Wernher von Braun, and had adopted the manned Mars landing program as a national commitment, an American manned spaceship might by now have already traveled to Mars. Not only that, but American technology might well have continued to maintain world leadership. America's anti-science disease, very simply, ended up killing the goose that laid the golden egg.

The post of Science Adviser to the White House that President Nixon had abolished in 1973 was resurrected by Congress in May, 1976, under President Ford. This decision was a result of pressures on Congress from the National Academy of

Sciences and other scientific groups, many of the organizations that now had science policy functions. However, the political prestige of the Science Adviser's position, having once been challenged and lost under the Nixon administration, was not easily restored.

President Jimmy Carter took office in January, 1977, and on March 18 he named Dr. Frank Press, Chairman of the Department of Earth and Planetary Sciences at the Massachusetts Institute of Technology, as his Science Adviser. However, the new nuclear energy policy announced by the President on April 7, a subject he took a personal interest in, certainly did not appear to reflect any advice from the Science Adviser or the staff in the Office of Science and Technology. In announcing the new policy at a press conference, President Carter said:

> We have recently seen India evolve an explosive device derived from a peaceful nuclear power plant, and we now feel that several other nations are on the verge of becoming nuclear explosive powers.
>
> We are now completing an extremely thorough review of our own nuclear power program. We have concluded that serious consequences can be derived from our own laxity in the handling of these materials and the spread of their use by other countries. And we believe that there is strong scientific and economic evidence that a time for a change has come.
>
> First of all, we will defer indefinitely the commercial reprocessing and recycling of the plutonium produced in U.S. nuclear power programs.
>
> Second, we will restructure our own U.S. breeder program to give greater priority to alternative designs of the breeder other than plutonium, and to defer the date when breeder reactors would be put into commercial use.

President Carter listed five more points in his new non-proliferation nuclear policy, but the first two items surprised nuclear energy experts everywhere. The two guidelines amounted to a total reversal of America's previous policy on nuclear energy.

In the reprocessing process, fuel that has been used by a nuclear power plant reactor is dissolved in chemicals like nitric acid to extract unburned uranium and plutonium that is created as a by-product in the nuclear fuel. The uranium and plutonium recovered can itself be used as fuel in a nuclear power plant. However, the recovered plutonium can also be used to build nuclear weapons. India, for example, used her own reprocessing facilities to recover plutonium from fuel used in nuclear power plants, and then utilized that plutonium in a nuclear explosive device. India's first test explosion was detonated in 1974.

President Carter believed that countries with the capacity to recover plutonium from its spent nuclear fuel would want to have nuclear weapons. Thus, it would be a good idea, he felt, to prohibit the use of plutonium as a fuel in nuclear reactors. The U.S. government was in the process of building a commercial nuclear fuel reprocessing plant at Barnwell, South Carolina, but it was decided to halt that project. This suspension was intended to set an example for the rest of the world. The "breeder reactor" mentioned in the second point of President Carter's policy was a new type of nuclear power plant then being enthusiastically developed in Great Britain, France,

West Germany, Japan, and other countries. It is often compared to a "magic oven," which one can fuel with ten logs to do one's cooking. Then, when one looks in the ashes afterwards, one finds eleven new logs which can in turn be used to heat the oven. That is, in effect, how the breeder reactor operates. To understand how it is possible to build a "magic" nuclear reactor like that, it is necessary to start by considering the nature of uranium. Natural uranium consists of a mixture of two "isotopes" of uranium, two different forms of the element, called Uranium 235 (U-235) and Uranium 238 (U-238). Of these, it is only U-235 that sustains nuclear fission in a reactor, while producing a large amount of heat. U-238, on the other hand, produces no heat in a reactor.

The proportion of these two isotopes in natural uranium is always 0.7 percent U-235 to 99.3 percent U-238. Thus, the ratio of usable U-235 to "unburnable" U-238 is extremely small. This natural mixture does not make a very efficient fuel, so the numerous nuclear power plants now operating in America and Japan use what is known as "enriched" uranium, which means that some of the U-238 has been removed so that the percentage of U-235 is increased. The "quotient of enrichment" is from 2 to 3 percent, meaning that the percentage of U-235 in the fuel mixture is between 2 and 3 percent instead of only 0.7 percent. The remaining 97 or 98 percent is still "unburnable" U-238, but inside an operating nuclear reactor, this U-238 is changed into a new kind of fuel, an isotope of the element plutonium known as Plutonium 239 (Pu-239).

The issue, however, is what percentage of the U-238 has turned into Pu-239. That ratio can be determined by comparing the amount of U-235 consumed in the reactor with the amount of newly-created Pu-239. In other words, when ten logs are burned in the oven, how many new logs are found in the ashes? Specialists call this ratio the "conversion ratio," and in the case of today's most widely used power reactors, the conversion ratio is 0.6. This means that when ten logs are burned, six new logs are found in the ashes. However, if one burns those six logs, the number of new logs will of course be only three and six-tenths logs. After another burning one will have only two and two-tenths logs. Thus, the supply keeps diminishing. With a power reactor whose conversion ratio is 0.6, if one starts with 100 kilograms of natural uranium, one can "burn" only 3.5 kilograms. At that point, the "kindling wood" is used up. To deal with this, the "breeder reactor" was devised. It is a reactor whose conversion ratio is greater than 1.0, so that if one fuels its "oven" with ten logs, one finds more than ten new logs lying in the ashes. By using this type of nuclear reactor, every bit of the natural uranium fuel can be "burned."

There was concern that if today's conventional (non-breeder) nuclear power reactors were to remain in use, they might exhaust the earth's uranium reserves in 30 years or so. But if breeder reactors are built, they will produce many times the amount of uranium they use. Conservatively assuming that such reactors can produce 20 kilograms of fuel for every kilogram of natural uranium originally inserted, this would extend the life of the world's uranium reserves to 20 times 30 years, or 600 years. The world's first nuclear reactor was built at the University of Chicago in 1942, through the efforts of the Italian-born physicist Enrico Fermi and other scientists. As early as 1945, Fermi said: "The country which first develops a breeder reactor will have a great competitive advantage in atomic energy."

For a country like Japan, for example, which has within its borders hardly any energy resources like coal or oil, and no deposits of high-grade uranium ore, the breeder reactor is a godsend. If the breeder reactor could be made practical, Japan would no longer need to purchase uranium fuel. Simply through the conversion into plutonium of uranium fuel already bought, Japan could easily obtain enough nuclear fuel to generate electricity for more than a century. For that reason, the nuclear fuel cycle of sending spent fuel to a reprocessing plant, extracting the plutonium, and using the latter as fuel in a breeder reactor, is seen as the ultimate pattern for nuclear power generation. Besides the United States, countries like France, Great Britain, West Germany, and Japan have all been keen to pursue this development.

Despite the virtues of this process, the government of the United States in 1977 suddenly declared a halt to its work on the nuclear fuel cycle. President Carter put it in these words:

> From my own experience, we have concluded that a viable and adequate economic nuclear program can be maintained without such reprocessing and recycling of plutonium.

With this announcement, he suspended construction work already in progress on the breeder reactor at Clinch River, Tennessee. He then referred to other countries:

> We are not trying to impose our will on those nations like Japan and France and Britain and Germany which already have reprocessing plants in operation. They have a special need that we don't have in that their supplies of petroleum products are not available.

Regarding Japan, however, President Carter's statement was not an accurate one. At the time he spoke, Japan's reprocessing plant had not yet gone into regular operation. Japan had spent seven years building a reprocessing facility at Tokai-mura, on the Pacific coast northwest of Tokyo, and had completed all required tests. The plant was on the verge of starting trial operation using actual spent nuclear fuel. Just at that juncture, President Carter appeared to be telling Japan, "Wait!" The timing was unfortunate. The Tokai-mura reprocessing plant itself had been built with French technical assistance. In order to ensure that the plutonium would not be diverted for use in nuclear weapons, the Japanese plant was subject to inspection by the International Atomic Energy Agency (IAEA). Accordingly, the American government should have had no basis for apprehension.

But there was just one problem: all of the spent nuclear fuel to be reprocessed there was enriched uranium that had originally been purchased from America. America could control the use of this material. According to the Atomic Energy Cooperation Agreement that had been signed by America and Japan, guidelines for the handling and disposition of nuclear fuel purchased from the United States were to be based on "joint decisions" by the American and Japanese governments. In order to arrive at such a joint decision, of course, the agreement of the American government was required. However, Washington seemed unwilling to say yes, thus forcing Japan to pour all her energies into negotiations that lasted from April to September, 1977. The policy had been made, now some kind of agreement had to be worked out.

In the end, a joint decision was reached, but from the standpoint of Japan, which desperately needed the breeder reactor, President Carter's new policy was truly a case of unwanted meddling. Perhaps America, with its comparative wealth of domestic oil and coal resources, had no need to force nuclear power development upon its people; but countries like Japan and France, which possessed few energy resources of their own, had only nuclear power to rely on. Hiromi Arisawa, president of the Japan Atomic Industrial Forum, Inc., has said of this issue: "The spread of nuclear weapons is a threat to world peace — but a scarcity of energy poses an even greater threat."

President Carter, nevertheless, when he was still a candidate, took an "anti-nuclear" stand, in hopes of gaining support from Ralph Nader and other environmental groups. Thus, if President Carter had not at least halted work on reprocessing plants and breeder reactors, he would have been accused of breaking faith with those who voted for him in the belief that he would oppose nuclear power development. President Carter appointed many environmentalists to important posts in his administration. Most of those officials were in what I call the anti-science camp. Consequently, during the era of President Carter, America still showed few signs of having recovered from the anti-science disease.

Since 1981, when President Reagan came to Washington, the American nuclear industry has further declined. Unable to come up with a balanced, safe but forward-looking nuclear policy, the U.S. government has surrendered leadership in this field to foreign nations. Japan, now freed of the Carter administration constraints on reprocessing, is moving ahead to commercialize nuclear reactor technologies. Inconsistent and poorly designed national policy has had its toll on America's science and technology leadership.

R & D Budgets Slashed

As America's anti-science epidemic deepened from field to field, spending on research and development began to dwindle, as did the number of persons actually engaged in research and development. The result was that research and development capability weakened, and America started to lose her ability to compete in the world market. Of course, American dollar spending on research and development is still the largest in the world. Figure 16 shows this spending in graphic form.

On May 8, 1981, I happened to turn my television set to the PBS program *Bill Moyers' Journal*, and found that the topic under discussion was military research spending. Responding to questions by Bill Moyers was Gordon Adams, an economist:

> MOYERS: What does defense spending, in your analysis, do to affect our innovative, productive, and competitive role in the world?
>
> ADAMS: I think this is the particular area of defense spending, as currently projected by the Reagan administration, where we're going to have to worry the most. You see, the Reagan administration is committed to a program, and I'm not saying it's an undesirable program, to revitalize the American economy. But I can't imagine a sector where you're going to succeed less in revitalizing the American economy than the defense sector.
>
> MOYERS: Why?

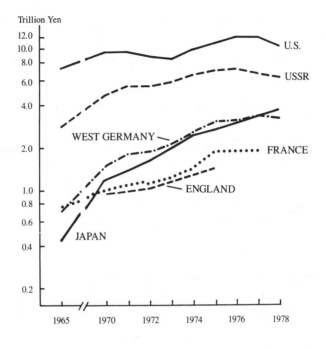

Figure 16   RESEARCH SPENDING IN SIX MAJOR COUNTRIES

ADAMS: What they're saying is that they're going to take a big package of government capital, that's going to be a third of the federal budget by 1986 —

MOYERS: The defense part of it.

ADAMS: The defense part of it. They're going to take a third of that budget, 65% this coming year, of all the federal commitment in research and development — that's everything that we as taxpayers fund to improve our technology, to improve our productive base — and they're going to put it into aerospace. That's 65% of our R & D, or a third of the federal budget in '80. Now, the problem is that you don't really get the spinoff in the commercial market from that kind of technology, that you get from directly investing in the kinds of technologies that will help us compete with the Germans and the Japanese. Now, look at the Germans and Japanese — they invest a far lower proportion of their total budget or their GNP or their R & D resources in defense and aerospace technology.

MOYERS: I think the Japanese spend about $80 to $90 per person per year on defense spending, and we spend $800 to $900.

ADAMS: That's right. It's actually a factor of about 6 to 1 between the Japanese and the United States.

MOYERS: And that's why we have to go abroad and buy many of their products.

ADAMS: They are in a position to put their research and technology exactly where it should be located, whereas we're distorting our capital investment in that sector, in directions that aren't really going to help us industrialize. You know, the person who really put his finger on this — this is kind of interesting — was Murray Weidenbaum, about 10 years ago.

MOYERS: He's the chairman of the Council of Economic Advisers.

ADAMS: He's now the chairman of the Council of Economic Advisers. Ten years ago he said, in an article that he wrote, that if you really want to revitalize the economy, about the worst place you can imagine doing it is putting your money into the defense sector. These are the least innovative firms, the least entrepreneurial firms, the ones with the least impact in the commercial economy. It's the wrong place to put your capital investment. So we're going to pay a price by investing so much of our resources now in a sector that has so little payoff where the American economy really needs it. It's a poor reindustrialization strategy.

Thus, even though the United States may spend nearly three times as much as Japan on research and development, there is some question as to whether the results of that research are actually worth three times as much. Total American research and development spending has continued to increase in nominal terms, as shown by the solid line in Figure 17. But if inflation is taken into account and expenditures are calculated in 1972 dollars, the broken line shows the resulting figures. Research and development expenditures in real terms are seen to have peaked in 1968, and then to have declined from 1969 to 1975. *That decline coincides with the period when the anti-science disease was at its worst in America.*

Next, let us look at Figure 18. This chart shows what percentage of each country's gross national product (GNP) corresponds to that country's expenditures on research and development. America reached its highest such ratio — about 3 percent — in 1964, and then the percentage steadily declined, until in 1979 (the last year shown) it reached a low of about 2.2 percent. This declining ratio is a result of the fact that spending on research and development failed to keep pace or catch up with a constantly increasing GNP. As can be seen from the chart, the USSR is on top, in terms of the ratio of research and development spending to GNP. The ratios for France and the United Kingdom have fallen like that of the United States, while those of West Germany and Japan have risen. The ratio for Japan, however, compared with

Billions

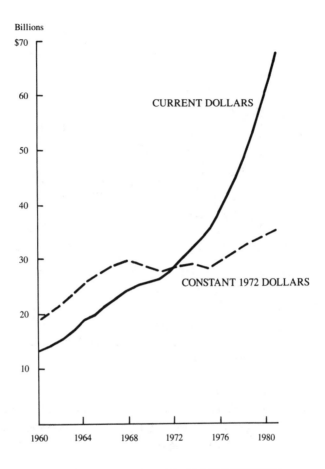

Figure 17   NATIONAL R & D EXPENDITURES

those of the other countries, is not at all large, leaving the possibility that it could still increase in the future. This chart indicates that the United States became stingy with research and development funding from around 1967. And it was not only spending on research and development that started to decline. The actual number of persons doing research also fell visibly, as is shown in Figure 19. The graph indicates year-by-year changes in the numbers of scientists and engineers engaged in research and development. From a glance, one can see that these numbers had been rising until 1969, but that from 1970 until 1973 they dropped off steeply. The leading factor in that drop-off was NASA's program reductions. In 1970, the year following *Apollo 11*'s successful moon landing, NASA carried out a large-scale personnel retrenchment. Because large numbers of NASA contracts with private companies also

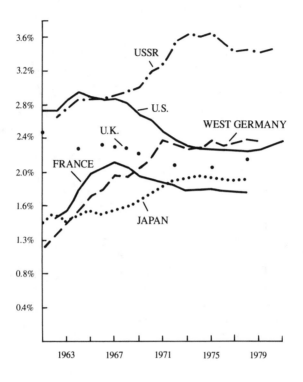

Figure 18 NATIONAL EXPENDITURES FOR PERFORMANCE
OF R & D AS A PERCENT OF GROSS NATIONAL
PRODUCT (GNP)

expired scientists were laid off by industry in steady streams. There were reports of rocket engineers who had been fired and were forced to become taxi drivers in order to survive.

The view is sometimes expressed in America that by spending money on such things as the space program, America ended up losing the "battle for trade." But when the space program was cut back, did the money thereby saved get transferred into other areas of science and technology? No, nothing of the sort. Research and development spending, and the number of research personnel, were arbitrarily slashed. When development of the supersonic transport was discontinued in 1971, the Boeing aircraft company laid off around 7,000 employees at one stroke. With one such event after another, the number of scientists and engineers employed in research and development steadily decreased. It was only in 1974 that the number began to turn upward once again. The years from 1970 to 1973 were truly an ordeal for America's scientists and engineers. Because the symptoms of America's anti-science

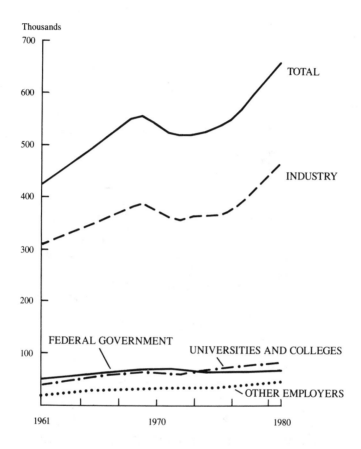

Figure 19   SCIENTISTS AND ENGINEERS EMPLOYED IN R & D BY SECTOR

disease were quite serious around that time, no one seemed willing to extend a helping hand to those scientists and engineers.

Other countries, however, were not so foolish. Figure 20 shows how many out of every 10,000 workers were persons engaged in research and development. America alone displays a drop-off on this chart, beginning in 1969. One predictable consequence of such a loss of research funds and personnel would be a slowdown in the pace of innovation and invention. Figure 21 charts the changes in the numbers of patents granted in the United States. The number reached a peak in 1971, then fell steadily after 1974. Particularly noticeable is the steep decline in patents awarded to Americans themselves, while at the same time the number of United States patents granted to foreign inventors climbed little by little. Today, about one-third of all American patents are obtained by foreigners. When the numbers of patent appli-

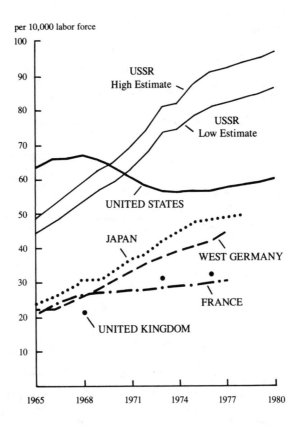

per 10,000 labor force

Figure 20   SCIENTISTS AND ENGINEERS ENGAGED IN R & D

cations filed in various countries are compared, the result is the graph in Figure 22. Of course, the sheer number of applications is no indication of the nature or importance of the inventions themselves, but it does suggest how zealously a given country is improving its technology. At the present time, Japan is the world leader in patent applications, largely because of the style of patent applications.

Michael Blommer, the executive director of the American Patent Law Association, has this to say about the patent situation:

> It's a symptom, a clear and unmistakable symptom, that America's fall-
> ing behind in new products and in products made with new processes.
>   While the United States pours money into an ailing giant like Chrysler,
> . . . in Japan and Germany they're pouring money into the most ad-
> vanced industries, computers, optical scanners. . .the things of the fu-
> ture.

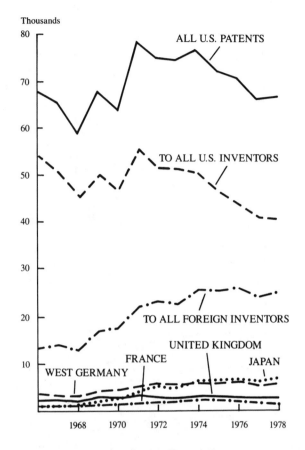

Figure 21   U.S. PATENTS GRANTED TO INVENTORS FROM
SELECTED COUNTRIES

Again, Donald Banner, director of the Intellectual Property Owners Association,
comments:

> Part of the problem may have been overconfidence in U.S. technology
> in the past. . . .We have not tried to stay up there. . . .We have just
> assumed that we would always be doing it better than anybody else.

America's balance of trade (value of U.S. exports minus value of U.S. imports) in
technology-intensive products, vis-à-vis several other countries, is shown in Figure
23. For the so-called developing countries, Canada, and Western Europe (minus
West Germany), the United States is still a net exporter of technological products. But

Figure 22   NUMBER OF PATENT APPLICATIONS IN MAJOR
INDUSTRIAL COUNTRIES

America has been a net "importer" of technology, as far as Japan is concerned, since about 1968. In other words, strictly in terms of technology-intensive products, America buys far more from Japan than she sells to Japan. Most recently, in fact, the United States has also become a net importer of such products from West Germany as well.

Education Neglected

The recent effort by state governments and local groups to improve the quality of American education after nearly two decades of neglect has proved a formidable task.

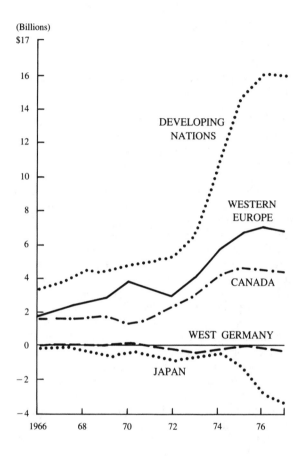

Figure 23   U.S. TRADE BALANCE WITH SELECTED NATIONS
FOR RESEARCH-INTENSIVE MANUFACTURED PRODUCTS
(1966–1977)

The quality and quantity of science instruction available to junior and high school
students declined, resulting eventually in a lower level of understanding of science
and technology by the general public. In October, 1980, the National Science
Foundation and the Department of Education compiled a report for President Carter
entitled *Science and Engineering Education for the 1980's and Beyond*. This
document analyzed in detail the deplorable state of scientific education in America's

schools. According to the report, high school students who aspire to become scientists and engineers are receiving a comparatively sound education, and are highly capable; but science education for those students who do not intend to become scientists and engineers has noticeably deteriorated. The report stated:

> Scientific and technical literacy is increasingly necessary in our society, but the number of our young people who graduate from high school and college with only the most rudimentary notions of science, mathematics, and technology portends trouble in the decades ahead.
> While students who plan scientific and engineering careers are receiving an adequate educational foundation, more students than ever before are dropping out of science and mathematics courses after the tenth grade, and this trend shows no signs of abating.

According to the report, only one out of six high school juniors or seniors elects to take courses in science or mathematics. Most of the students belonging to that "one out of six" category are planning to become scientists, engineers, or physicians. This means that students planning careers in non-scientific fields have not taken courses in either mathematics or science by the time they reach their junior year in high school. Since 1980 a number of reports on the decline of American elementary and secondary education have appeared. In 1984 the basic issues were discussed in a number of important studies. On June 7, 1984 the *New York Times* reported on a study by the National Academy of Sciences that pointed out that the annual number of American students obtaining doctorates in mathematics dropped by more than 50 percent from 1968 to 1982. Much of the decline was attributed to the failure of federal funding and the lack of national priorities for math and science education.

How did this situation come about? The 1980 report suggests two reasons. First, there is the matter of coursework requirements for high school graduation. There are about 17,000 high school districts in the United States, each permitted, within broad limits set by state boards of education, to decide its own curriculum, standards, and requirements. Since 1970, many districts have relaxed their high school graduation requirements, until today only one-third of all high school districts require graduates to take more than one year of mathematics or science. By way of contrast, fully three-fourths of all high school districts require students to take more than one year of social studies in order to graduate. This comparative neglect of mathematics and science may well be one of the effects of the wave of anti-science sentiment that swept America.

A second reason for the relative decline in the number of students electing science and mathematics courses in high school, according to the 1980 report, was "the reduction of standards for admission and retention by some colleges and universities, in response to increased competition for students." This has tended to exert an adverse effect upon academic standards at the high school level, with students tending to shun the more demanding science and mathematics courses. The decreased enrollment in these courses in high schools and colleges resulted in a lower level of understanding of science among the American people as a whole. The report describes the problem in these words:

The role of science and technology is increasing throughout our society. In business, in government, in the military, in occupations and professions where it never before intruded, science is becoming a key to success. Yet our educational system does not now provide such understanding.

Such a decline in the general public's scientific understanding is a major political and social handicap. David Savage, associate editor of the weekly magazine *Education USA*, writes: "How, after all, can voters intelligently choose between a pro- and an anti-nuclear candidate if they cannot make sense of how nuclear power works?" The problem is not only in the voting booth, and the National Science Foundation report cites two examples. One is that according to reports from some American manufacturers, over the past decade the average worker's basic comprehension of science and mathematics has apparently declined to a point where more time is now required to manufacture a given product in certain highly technical areas. The other example is that of the American armed forces. Weapons have become increasingly sophisticated, but because of the public's declining general awareness of science and mathematics, it is more and more difficult to find officers and non-commissioned officers who are qualified to receive training to operate and maintain the revolutionary but complex new military hardware.

In the competition for international trade, as well, America is inevitably placed at a disadvantage. The report states:

> The declining emphasis on science and mathematics in our school systems is in marked contrast to other industrialized countries. Japan, Germany, and the Soviet Union all provide rigorous training in science and mathematics for all their citizens. We fear a loss of our competitive edge.

The report includes a table showing what percentage of young people of the same age group in four major countries are graduates of college-level engineering curricula. This chart, reproduced here as Table 2, indicates that Japan has the highest percentage, 4.2 percent, of engineering graduates. West Germany is in second place, with 2.3 percent, while the United States with only 1.6 percent, is below even the United Kingdom.

Table 2

	Share of World Trade		Engineering Graduates as a Proportion of Relevant Age Group
	1963	1977	1977-78
United Kingdom	15%	9%	1.7%
W. Germany	20%	21%	2.3%
Japan	8%	15%	4.2%
United States	21%	16%	1.6%

In recent years, Japan has produced larger absolute numbers of engineering graduates than the United States, and this despite the fact that America's total population is twice that of Japan. Twenty percent of Japan's university graduates come out of departments of engineering, compared to a mere 5 percent in America. The report points out that in Japan, moreover, the educational level of workers in manufacturing industries is high, and that some 50 percent of company directors have engineering qualifications.

The report also describes in detail the Japanese middle school and senior high school curriculum. About 25 percent of class instructional time in grades seven through nine (middle school) is devoted to mathematics and science. Thus, virtually every citizen has studied these subjects. Geometry is studied every year in grades seven through nine, with trigonometry required in ninth grade. In senior high school, the curriculum includes differential and integral calculus, probability theory, and statistics. The report declares that "mathematics instruction has a more rapid pace in Japan than in the United States, and a much higher proportion of students take the more advanced courses." Japan's success in teaching mathematics has been established in international tests. The mathematical abilities of 13-year-old students in twelve countries, including the United States, Japan, and several European countries, were compared through the use of standardized tests. The Japanese students' average scores were the highest of the twelve.

In attempting to emphasize to the President the critical state of America's science education, the National Science Foundation report, in my view, tends to overpraise science education in Japan, West Germany, and the USSR. For example, according to the report, 65 percent of Japanese university graduates major in the natural sciences. In America, on the other hand, says the report, only 30 percent do so. It is true that in Japan, science majors outnumber humanities majors at national (government run) universities like the University of Tokyo, but Japan's privately-endowed universities concentrate their enrollments primarily in the non-scientific departments, which do not require expensive laboratories and research facilities. If these non-governmental universities are included in the statistical averages, then Japan also has only around 30 percent of its university graduates majoring in the natural sciences.

As the report points out, Japanese science instruction, starting in seventh grade, is extremely comprehensive in content, and taught at a level of difficulty perhaps unsurpassed in any other country. But because so much is taught, the fact is that fairly high percentages of students are unable to keep up with the pace of instruction, and fall behind or drop out. Moreover, Japanese educators are often accused of merely trying to stuff the heads of children with bits and pieces of data. The most important aim of science education, to many people, should be to make the student fully aware of the scientific way of thinking. Too little attention is given to reasoning and logic in science in Japan. As a result, even among scientists and engineers, there are all too many who are incapable of scientific thought and judgement outside their own particular fields of specialization.

In Japan, moreover, when the anti-science disease was also at its worst (between 1970 and 1975) university engineering departments were unable to attract the outstanding students. A professor of chemical engineering at one Japanese university lamented to me: "I don't know whether it's because the chemical industry is being blamed as the main cause of pollution, or what, but lately the only students who want to major in my department are ones with poor grades. I'm really worried about the future." Examined at close range, Japanese scientific and engineering education is also full of defects. In America, however, as the report points out, the situation is probably far worse. The deterioration of educational standards is a tragedy that will have serious and long-lasting effects. William G. Aldridge, executive director of the National Science Teachers' Association, warns:

It might well be that an inadequately prepared citizenry in a highly technological world could help reduce the United States to third-rate power status well before the year 2000.

VIII
RESTORING AMERICA'S TRUST IN SCIENCE

America's anti-science disease raged for more than 15 years, beginning in 1965, but from 1981 to the present American newspapers, television and other mass media have come to realize the dangers of those anti-science attitudes. A good example of this was the media's coverage of the first mission of the space shuttle *Columbia*. The reporting in the *New York Times* was particularly extensive and forceful; one article about *Columbia* by the paper's science reporter, John Noble Wilford, was so enthusiastic that it seemed as if a pro-science campaign was underway. As we have seen in Chapter I, Wilford wrote an article entitled "Space and the American Vision," which appeared in the *Times* about one week before *Columbia* was launched. In this article, Wilford deplored the decline of American science and technology and expressed the hope that a rebirth was imminent. He wrote:

> The idea of sending a human being hurtling off into an alien sky is no longer so frightening or novel. Nor will there be the same sense of history that attended the launching of *Apollo 11* to the moon. But circumstances have conspired to burden the flight with an importance beyond its immediate objectives.
>
> This time it is almost as though we are counting on *Columbia* to show us that we are the equal of ourselves, the equal of our own traditions, of our own images of present and future. It is as though we are looking to space to help us resolve some of the inherent conflicts that seem to exist for us at home in the 1980's. Here we are, a nation that has for decades led the world in technological development, that still leads the world in national production, but that has slipped back significantly in some areas relative to some nations.
>
> It is as though the fires of Canaveral will be testing anew not only a traditional commitment to technology and science, but also, really, a deeper spirit of adventure and frontiersmanship.

Wilford quoted Joseph W. Loftus, Jr., chief of technical planning at the Johnson Space Center in Houston, as saying: "The advent of the shuttles could do for the economic development of space what the transcontinental railroad did for the American West." The Apollo spacecraft, Wilford wrote, was still at the "covered wagon" stage of space travel. As long as the only way to cross the plains and mountains was by wagon, the economic development of the West remained a slow process. Only with the building of the transcontinental railroad did rapid growth become possible. In a similar manner, said Wilford, the space shuttle will rapidly accelerate the development of space.

Most people are now familiar with the capabilities of the space shuttle. Unlike the Apollo spacecraft used in the lunar landing mission, the space shuttle is a winged vehicle, which coasts down through the atmosphere like a glider and lands on the same type of runway used by a jet aircraft. After only two weeks or so of preparation, the same space shuttle can be launched again into orbit. In this manner, a single space shuttle can make a hundred or more trips into space and back again. For carrying loads, the space shuttle has a cargo bay 15 feet in diameter and nearly 60 feet in length. This compartment can also carry a large laboratory for conducting experiments that are possible only in orbit, like the recent combining of a light and a heavy metal into a new alloy. Another such space experiment involves the manufacture of glass containing virtually no impurities. When glass is melted on the earth's surface in the usual type of crucible, impurities from the crucible wall itself inevitably become mixed with the molten glass. The interior of a space shuttle, however, is in a weightless state, allowing glass to be melted without being held in a crucible — for molten glass, in zero gravity, will float in space as a spherical globule. If this glass globule is skillfully manipulated, it can be drawn out into extremely fine glass fibers.

The space shuttle tends to inspire expectations like these, and John Wilford was not the only journalist to write in homage to the launching of *Columbia*. The columnist Henry Fairlie of the *Washington Post* emphasized the significance of the launching by likening it to the embarkation of Columbus from Spain on August 3, 1492. Instead of discussing the material gains to be obtained from "Made In Space" manufacturing, Fairlie noted the importance of the astronomical observatory satellite, which scientists plan to boost into orbit aboard the space shuttle. The attitude of America's press, radio, and television toward the first space shuttle was, in general, highly favorable. Hardly any newspapers, for example, carried editorial comment criticizing the launching of *Columbia*. Journalists seemed to be trying to justify the project even if it should end in failure. Henry Fairlie wrote:

> And if the shuttle does indeed nosedive to earth two weeks from now we ought not to smirk at the disaster. We ought to consider what would be one of its causes. The space program depends on the long-range development of the most sophisticated technology. It has for years now been subject to erratic funding and arbitrary cuts in its budget. If the shuttle crashes, it will be our fault.

When *Apollo 11* made the first lunar landing in July, 1969, American newspapers and magazines carried a considerable amount of critical commentary. But with the first space shuttle, attacks of this kind virtually disappeared from the press. Nor did any demonstrators show up at the Kennedy or Johnson Space Centers. Not one placard demanding "Milk For Babies, Not Toys For Space" was to be seen.

America's intellectuals vigorously opposed development of the supersonic transport — it would make as much noise as fifty jumbo jets, they asserted, and its sonic boom would break windows, and the water vapor in its exhaust gases would destroy the ozone layer in the atmosphere. But the noise of the space shuttle's launching rockets was probably far louder than any SST — loud enough to rattle the roof of the press boxes 3.5 miles away from the launch pad. Moreover, the space shuttle's main engines burn liquefied oxygen and hydrogen, producing an exhaust gas consisting entirely of water vapor — which would seem to be as harmful to the ozone layer as the

SST's exhaust. Finally, the *Columbia*, when landing at Edwards Air Force Base in California after returning from orbit, emitted two sonic booms. And according to NASA's plans, space shuttle flights will be launched at the rate of about one per week by the mid-1980's.

If *Columbia* had been ready to make her maiden flight in 1971, what would have been the reaction? Would the newspaper and television commentaries have been as laudatory as they were in 1981? America's press once lent its support to the anti-SST campaign, but today's newspapers are full of goodwill toward the space shuttle. It seems to me that after 15 years of deluding itself, American journalism has finally awakened to the importance of science and technology, and has begun to recognize the dangers of the earlier anti-science attitudes. The *New York Times* journalist William Stockton wrote the following in an article headed "The Technology Race":

> After decades of global supremacy, American technology is faltering at precisely the wrong point in history — a time of great scientific innovation throughout the world. Getting the United States back on the track is challenging the nation's scientific and industrial communities.

America, concludes Stockton, must "roll up the national sleeves and get to work." In 1957, when the Soviet Union launched Sputnik 1, the American people, jolted out of their complacency, rolled up their sleeves and went right to work. They successfully achieved the first landing by human beings on the surface of the moon, overtaking the Soviet Union in the race for space. Today, the threat of Japanese technological superiority is proving to be a second Sputnik in its effect upon the people of the United States. The effect, however, has been slow to take on a Sputnik-type character. Even the 1984 elections gave little attention to the issue.

Contract for Survival

Japan never suffered as badly as the United States from doubts about science. Japan came down with the illness five years later than America and awoke to the perils of the disease four or five years before America. The period during which Japan suffered most severely from the effects of anti-science lasted, at most, six or seven years (1970-1977). This difference of nearly a decade gave Japan the opportunity it needed to catch the United States in many scientific areas.

Why did Japan recognize the dangers of anti-science so much sooner? Why was Japan ahead of America in taking steps to treat and cure the disease? The initial impetus to Japan came from the first of the so-called "oil crises," which developed in the autumn of 1973. When the Middle East war broke out in October, the Arab countries decided to use their oil as a strategic weapon by reducing oil exports to other countries. Simultaneously, they raised the price of oil by about 70 percent. It is probably hard for most Americans to comprehend the shock that the Japanese people felt at their only source of oil being threatened. At the time of the 1973 oil cutoff, an estimated 83.5 percent of Japan's total energy was imported and nearly 100 percent of its oil came from abroad. The cutoff of oil was a threat to the whole economy.

The United States, on the other hand, supplied from within its own borders fully 90 percent of the energy it was using in 1973 — only 10 percent of its energy needs had to be met by imports from other countries. Consequently, the Arab decision to cut oil exports was not nearly as much of a blow to American total interests. But in Japan, the public was unable to remain calm. The first sign of this, silly as it seems when looking back now, was the "toilet paper hysteria." If no oil were to arrive in Japan from the Arab countries, paper manufacturing would become impossible, and supplies of toilet paper would be exhausted. Japanese housewives, who must have envisioned desperate times ahead in the family bathroom, went on a toilet paper hoarding spree. Supermarket shelves stacked high with rolls of toilet paper were swept bare in hours as hordes of housewives rushed to stock up before supplies ran out. On November 2, 1973, the toilet paper stampede became so violent that injury resulted. At a consumers' cooperative store in the city of Amagasaki, near Kobe, a mob of 200 housewives shoved their way inside as the store's doors were unlocked at the opening of business. Several women were pushed from behind, fell down, and were trampled by others in the mass of people, as the crowd surged toward the shelves of toilet paper. One 83-year-old woman who fainted and fell under the feet of the crowd suffered a broken right leg.

But the hysteria did not stop with the run on toilet paper. Housewives also began hoarding synthetic detergents (made from petroleum), and even articles not made from oil, like matches, sugar, and salt. The Japanese government announced oil conservation measures. Illuminated advertising display signs were extinguished; bars and night clubs had to close at midnight; automobile owners were supposed to refrain from making long trips; and gas stations remained closed all day on Sundays. What alarmed people more than anything else was the steady upward climb of the price of oil. At the beginning of 1973, Middle East crude oil had sold for only $1.80 per barrel, but by the end of that year, the price had soared to $11 or even $17 per barrel. One year had seen a fivefold increase in oil prices. This drained the economy.

Japan uses exports of manufactured products to support the living standard of its people. The Arab countries can sell the oil that lies beneath their deserts, and the United States has vast plains producing rich harvests that can be sold abroad. But Japan has no natural resources, and not enough land for large-scale agriculture. Japan's strategy is to purchase raw materials from other countries, turn them into high-technology manufactured products, and export those products. With its export earnings, Japan is able to buy from other countries the oil, food, and raw materials needed by its people and for its industries. If those imports of oil and raw materials are cut off, or become too expensive, Japan's "trade cycle" is broken.

With the petroleum crisis in 1973, many Japanese had thoughts like this: "If the price of oil keeps rising, we will have to invent techniques to conserve oil and other forms of energy in all manufacturing processes." "If our oil supplies are going to be uncertain, we need to develop reliable energy sources that can replace oil." "If the prices of commodities like oil continue to increase, so that we have to keep paying more and more to foreign countries for our imports, then we've got to come up with high-technology-intensive products, with even greater added value than today, and export them in order to earn the foreign currency we need. Otherwise, Japan will end up going bankrupt." The Japanese business and government leadership turned to technology for solutions.

Japan's case of the anti-science disease was not completely cured by the shock treatment of the oil crisis. As I have noted previously, the Japanese people have a habit of blindly imitating whatever Americans do. When Ralph Nader became active in America, activists who might be called "Naderites" appeared in Japan as well.

They imitated precisely everything Nader said and did. A leader of the Japanese consumer movement, who was nicknamed "the Ralph Nader of Japan" by the mass media, was overjoyed at the honor. It was after the oil crisis, in 1974, that Ralph Nader's movement opposing nuclear power in America reached its peak. In Japan as well, therefore, an anti-nuclear power movement flourished at the same time. The agitation against the nuclear-powered ship *Mutsu* lasted from the spring through the fall of 1974.

In Japan in 1974, the anti-science disease had started to subside a little, but now reinfection by American anti-science attitudes again caused a relapse. During this recurrence of the anti-science disease, the Tokyo newspaper *Asahi Shimbun* began publishing (July 1976) a series of articles in support of nuclear power, written by a woman reporter on the paper's science staff. At the time, any newspaper that dared to publish an article favorable toward nuclear power would soon find anti-nuclear activists bursting into its offices to register loud and even violent protests. Thus, to write and publish a whole series of articles advocating nuclear power took a considerable amount of courage on the part of both the reporter and the newspaper.

The social critic Soichiro Tahara, in his book *The Contract For Survival*, wrote the following account:

> The movement seeking the extinction of nuclear power had spread like wildfire, but suddenly it seemed to weaken, and before long it changed its direction. For example, most of the newspapers and magazines that had formerly devoted large amounts of space to articles and interviews about the leaders and the students in the movement gradually changed their attitude to one of virtually urging the development of nuclear power. When did this change of direction begin? Most likely, the turning point, the catalyst, as it were, was a series of 48 articles published in the *Asahi Shimbun* between July 13 and September 5, 1976, under the title "Nuclear Fuel: From Prospecting to Waste Disposal," written by Yukiko Okuma, a member of the *Asahi's* science department. The newspaper subsequently published the series in book form. . . .Many readers were highly critical of the series, and numerous citizens' anti-nuclear groups went to the newspaper's offices and protested to Yukiko Okuma, so great was the impact of this series upon the opponents of nuclear power.

Yukiko Okuma had visited Japan's nuclear power facilities, and had taken a close and careful look at each one. She had descended into the depths of a uranium mine, and had inspected a factory where natural uranium is enriched. Her visit to the reactor room of one nuclear power station is said to have caused quite a stir. It seems that in order to enter a reactor room, a visitor is required to remove all street clothes, strip totally naked, and put on special bright-red clothing. However, the particular nuclear power plant Okuma was visiting on this occasion had only a men's dressing room for this purpose.

The plant manager told her: "You say you don't mind if anyone else watches, but still. . .! The cover of the reactor is open now, for the regularly-scheduled inspection, and if the workmen see you naked, it might cause an accident!" When Okuma persisted, the manager finally arranged for a vinyl sheet to be draped across one

corner of the men's dressing room, to create a "women's changing room," and
Okuma was at last able to put on the prescribed uniform and enter the reactor room.
She also inspected a laboratory studying nuclear waste disposal, and visited a
reprocessing facility, where plutonium is extracted from spent nuclear fuel. Using the
information, observations, and interviews she had so carefully collected, Yukiko
Okuma wrote her series of articles. Eventually the newspaper series was gathered into
a book, and in an "Afterword" to that volume the author wrote:

> At the time I was writing these articles for my newspaper, some mem-
> bers of the consumer movement and the political reform movement
> were busily promoting a "Ban All Nuclear Power" campaign. Among
> journalists, the prevailing sentiment seemed to be that opposition to nu-
> clear power was virtually a social imperative. I had read the writings of
> a great many opponents of nuclear power, and when I actually met
> them, I was surprised to discover that these anti-nuclear people had no
> accurate knowledge about nuclear fuel, or the effects of nuclear radi-
> ation upon the human body. Most of them appeared to base their anti-
> nuclear views on little more than secondhand quotations from American
> anti-nuclear pamphlets. I wrote my articles, and this book, in the hope
> that members of the public, including opponents of nuclear power,
> could gain some practical basic knowledge about nuclear fuel, and
> thereby become better able to make calm, rational judgments on matters
> involving nuclear power. After several years of gathering information
> on nuclear fuel, and studying the issues involved, I myself have finally
> reached this conclusion: A country as poor in natural resources as Japan
> must inevitably decide to draw energy from nuclear fuel.

In an interview with Soichiro Tahara, Yukiko Okuma explained in greater detail
her motives for writing the series of articles:

> Actually, as human beings, I liked the people in the anti-nuclear move-
> ment. But some of them — I don't mean to say all of them by any
> means — had a strong tendency to exaggerate, saying things that
> weren't true, knowing very well that they were lies, and to use threat-
> ening language, just to stir up anti-nuclear emotions. The people who
> were determined to bury nuclear power at any cost seemed to have the
> loudest voices, and I felt there was a danger in that. I mean, if you ask,
> "Can't we give up nuclear power in favor of some other program, or
> find some other kind of energy to replace nuclear power," the answer is
> no. I had been studying solar power, wind power, geothermal power,
> and so on for ten years, but I came to the conclusion that at least for
> the next ten or twenty years, we have no choice but to depend on nu-
> clear power. It's very irresponsible to arbitrarily demand an end to nu-
> clear power, when there's no alternative in sight.

Some people, remarked Tahara, seem to be saying that it might be all right just to
lower our present standard of living a little. Okuma replied:

There have been people who say things like that, but I simply can't believe that mankind would be better off if we slashed our energy consumption and went "back to nature." If society as a whole were to become poorer, what would actually happen is that inequalities in living standards would become greater, and it would be the people at the lowest levels who would suffer most. In their hearts, even the people who say we should simply reduce our standard of living and go back to nature don't really want to lower their own standard of living, and they aren't really expecting such a thing to actually happen. They're just talking trivialities. So I call that irresponsible and insincere. By comparison, the engineers at nuclear power plants are honest and serious-minded. They're really devoted to their work, and their technology is tremendously advanced. When I was gathering my material, I came to realize that. I can trust those engineers, and I can trust nuclear power if they are running it. But those engineers are becoming discouraged. Their work is being denounced as anti-social, and the devil's work, and so on, and they even get beaten up. Every newspaper or magazine they see is concentrating its fire on them, and finally, even their wives and children treat them coldly. So they're put in a terrible bind. Anybody who is constantly placed in that kind of situation tends to lose his desire to keep going, and will probably try to escape, eventually. And in fact, we had started to see signs of that. I felt that was a real danger. Up to then, I had thought that by adopting a severely critical attitude toward nuclear power, we could encourage it to develop in a healthier manner — but if we are too arrogant and self-righteous, then not only the nuclear engineers but nuclear power itself will be destroyed. And that would really be a disaster.

After she wrote her newspaper series, Yukiko Okuma was denounced by anti-nuclear intellectuals and "progressive" cultural personalities. Articles attacking her appeared in many newspapers and magazines. Nevertheless, her series of articles was a major factor in Japan's recovery from the anti-science disease. Subsequently, other members of the *Asahi Shimbun's* science reporting staff continued this effort to cure Japan of anti-science thinking. In February, 1977, for example, a campaign opposing "irradiated potatoes" sprang up in Japan, attacking the practice of exposing potatoes to nuclear radiation to prevent them from sprouting, thus permitting long-term storage. Japanese testing of this technique had begun in Hokkaido in 1973, and irradiated potatoes had appeared on the Tokyo market in the spring of the following year. By the autumn of 1976, however, consumer groups had begun to voice doubts about the safety of these potatoes.

At the time, the prevailing wisdom seemed to be, "the voice of a consumer group is the voice of an almighty force." And so in February, 1977, the Tokyo metropolitan government demanded that the Tokyo Fruit and Vegetable Dealers' Association notify the consumer whenever nuclear-irradiated potatoes were sold "loose" or unpackaged. Later, in March, the Japan School Lunch Association adopted a policy of not utilizing irradiated potatoes in school lunch programs. But in the *Asahi Shimbun* on March 14, 1977, there appeared an article by Eiji Mikado, of the newspaper's science department, headlined, "Repeated Tests Confirm No Danger of Radioactive Contamination." The article explained in detail how a large number of experiments and studies had repeatedly demonstrated the safety of irradiated potatoes. Mikado wrote, "Experts are unanimous in assuring the public that irradiated potatoes are safe, no matter how many are eaten." With this article, the anti-potato campaign did not flare up like the 1975 movement opposing lysine-enriched bread in school lunches (described in Chapter V). Instead, the opposition quickly disappeared. Irradiated potatoes continued to roll out of the test facility in Hokkaido.

In the summer of that same year (1977), great excitement was aroused by news reports that a Japanese vessel fishing in waters near New Zealand had hauled up in its nets the dead body of a "monster." Before the "monster" had even been examined by scientists, it was popularly dubbed "New Nessie," amid widespread speculation in the press that dinosaurs of the Mesozoic Era might still survive today. As with the Loch Ness Monster, un-scientific "common knowledge" seemed deeply-rooted in the minds of the general public. But an *Asahi Shimbun* science reporter, Akihito Oka, made a scientific study of the "monster," including a chemical analysis by scientists of bristles and other parts of the remains brought back by the Japanese fishermen. As a result of his research, Oka concluded that the supposed sea-monster was actually the corpse of a basking shark. Had it not been for the scientific efforts of reporters like Oka, perhaps an elaborate expedition might have set out by ship from Japan for the waters off New Zealand, to search in vain for the "New Nessie."

Japanese were leaning in the same direction as the Americans on science but the power of the economic elite and the authority of the media helped head off the massive failures in science policy that were evident in the United States.

Media Responsibilities

In contemporary society, the press, television, and the other mass media constitute an impressive power structure. The media possess the power to pull much of the public in a particular direction, and can exert a major influence upon political events. When newspapers and television fail to use this power wisely, however, the public can be misled. Since early 1981, America's newspapers have increasingly emphasized the importance of science and technology. Four or five years behind Japan's newspapers, the American press has finally begun to take steps to counter anti-science attitudes. Until 1981 America's media had been helping to aggravate this anti-science disease. By giving extensive coverage and publicity to the forces opposing science and technology, the media had been encouraging anti-scientific sentiments among the general public. Opposition campaigns and consumer movements alike would in general have been ineffectual without exposure in, and support from, the mass media.

At the beginning of Chapter V, I described how the anti-SST movement was started by Dr. William A. Shurcliff, working only with two members of his own family. Shurcliff and his tiny staff applied themselves intently to learning how to use the mass media most effectively. Shurcliff's principal strategies consisted of sending letters to the media, offering them articles and news stories, and running advertisements containing anti-SST propaganda. Other such campaigns have often used similar techniques. In order to obtain newspaper and television coverage, advocacy groups will hold press conferences, sponsor discussions, invite prominent figures from abroad, besiege government offices, and stage demonstrations. The print and electronic media treat such activities of the movement's core group as news events, faithfully reporting every assertion of the activists. The media seldom bother to find out whether the movement's statements are scientifically accurate, or to consider whether the movement will benefit the health or daily lives of the general public. Even when statements by movement activists contain falsehoods or exaggerations, they frequently appear verbatim in the press and on television. A journalist who afterwards discovers a statement to have been false, will rarely bother to write a story calling attention to the untruth. Instead, the journalist may even offer an excuse like this: "Well, that's what they said at the time. The fact is that they said it, so what's wrong with just reporting what they said?"

But actually, in such a case, unless the reporter writes, "That is what they said — but they were mistaken," the reporter is helping to propagate an untruth among the public. The reporter becomes an accomplice to a lie. When it comes to matters involving science and technology, the possibility of the journalist becoming a partner in an untruth, of reporting inaccurate information unchallenged, is much greater than in most other types of stories. That is because it often takes a long time to determine whether or not a given statement is scientifically correct. Newspaper and television reporters are constantly pressed by tight deadlines, so they are unable to wait for scientific inquiries to be completed. The train invariably has to leave before everyone is on board, as it were, and the probability of making a mistake is inevitably high.

Also, newspaper companies and television stations are not fully equipped to deal with the age of science and technology. Because most newspaper and television reporters are writers by profession, many of them have been educated primarily in the humanities or social sciences. Today, about 30 percent of all college graduates have majored in the natural sciences, but perhaps no more than one percent of all newspaper and television reporters were natural science majors. Of course, among the reporters who did not concentrate in the natural sciences can be found many fine individuals who possess a knowledge of science and are capable of thinking in a scientific manner, and they can produce comprehensible, high-quality science reporting. Many reporters coming from non-scientific academic backgrounds are indifferent to matters scientific and technological, and unwilling to inform themselves on such subjects. To combat these problems and to encourage accurate science reporting by making it easier for reporters to contact reputable scientists, the Scientists' Institute for Public Information recently established the Media Resource Service. Over 15,000 scientists and engineers participate in the Service to quickly connect journalists with people who may be able to clarify the issues.

Consequently, media reporting of a story like the accident at the Three Mile Island nuclear power plant often presents a spectacle that is pitiful to behold. As one false report follows another, the media coverage becomes more and more emotional, hysterical, and sensationalistic. The accident at Three Mile Island occurred at 4:36 A.M. on March 28, 1979. The technical details of the accident can be omitted here, but an accumulation of human errors played a major role: a valve that was supposed to be kept open was left closed by repairmen, and safety systems that went into operation during the course of the incident were deliberately turned off by plant operators on duty, as they tried to cope with the rapid series of events. The plant personnel at Three Mile Island were careless and inadequately trained, and demonstrated poor judgement, but the officials of the Nuclear Regulatory Commission (NRC) in Washington showed far worse behavior. Misinterpreting reports coming from the accident site, they gave the governor of Pennsylvania what amounted to an evacuation recommendation, and told newspaper reporters about a "bubble" of hydrogen gas inside the reactor — a bubble that actually posed no danger of explosion. Press, radio, and television reporting of the accident was exaggerated all out of proportion, because the NCR, the primary information source, was unable to supply the media with accurate and reliable information. However, the press corps itself was not without its own problems.

At the time of the Three Mile Island accident, I was at the Tokyo offices of my newspaper, and read all of the dispatches that came in on the various wire services. The reports were basically erroneous from the very start. A dispatch from Washington referred to "the undeniable possibility that 500 employees on the night shift at the nuclear plant have been contaminated by radioactive material." Reading this cable, we were incredulous: there must be some sort of mistake, we thought. It totally defied common sense to talk about "500 employees" being on the night shift of a nuclear power plant in normal operation. A nuclear plant is highly automated, and there would never be anything like 500 workers packed inside such a facility. Although we felt skeptical about the reports, we were under deadline pressures, and so without making any inquiries to the American offices of the wire services, we allowed their dispatches to be published unaltered in the *Asahi Shimbun*. Naturally, the Three Mile Island accident made the top of the front page, under a headline reading "500 Workers Feared Contaminated." We found out afterwards that at the time the accident occurred, the actual number of employees at the plant was only 60 or so.

On April 4, we received another wire service report stating that "twelve times the permissible dose of radioactive iodine has been detected in cow's milk." The amount that had been detected was 40 picocuries per liter of milk. We were dubious that this tiny amount could be twelve times the permissible dose, so we sent an inquiry back to the wire service's main office. The latter simply replied that their information came from an NRC announcement. As it later turned out, the NRC's figure for the maximum permissible dose was wrong, and the radioactive iodine in the milk was actually less than one-three-hundredth of the permissible dose. The Three Mile Island events were covered by 40 television cameras on the scene, and the total number of journalists at the site is said to have exceeded 300. A few of them were knowledgeable about nuclear power, but most had not the slightest familiarity with the subject. Moreover, the government and industry spokespeople who provided information to these journalists at press conferences also tended to be confused. A sense of panic seemed to envelop the Three Mile Island plant and its vicinity. The accident caused major economic damage to the power company. In fact, that was almost certainly the greatest single harm that resulted from the accident. But the people who lived near the Three Mile Island nuclear plant suffered virtually no ill effects. The quantity of radioactive material that escaped from the reactor was minute. The amount of radiation to which residents of surrounding areas were exposed was, on the average, no more than 1.5 millirems — less than one-sixtieth of the amount of natural radiation to which they were exposed in one year.

The Three Mile Island incident, rather than revealing the dangers of nuclear power plants, actually serves to demonstrate how much attention is devoted to safety in nuclear design and construction. Despite the fact that inexperienced operators and repair personnel committed a considerable number of human errors, not one person was killed or injured and not one resident of the vicinity was exposed to enough radiation to compel concern for health. These facts were accurately reported by my newspaper, the *Asahi Shimbun*, after the initial excitement of the incident had died down. For example, on April 8, 1979, the paper printed an article headlined "Measurements Show Decline in Radioactive Contamination: No Danger of Cancer Increase." On August 22, the *Asahi Shimbun* carried a detailed sumary of the report submitted by the President's commission on the accident at Three Mile Island, and made an effort to correct the erroneous impressions received by many readers at the time of the accident.

The Presidential Commission discussed the media's coverage of the accident, and made these comments:

> [One] severe problem was that even personnel representing the major national news media often did not have sufficient scientific and engineering background to understand thoroughly what they heard, and did not have available to them people to explain the information. This problem was most serious in the reporting of the various releases of radiation and the explanation of the severity (or lack of severity) of these releases. Many of the stories were so garbled as to make them useless as a source of information.

> [Journalists] experienced difficulty interpreting language expressing the probability of such events as a meltdown or a hydrogen explosion; this was made even more difficult when the sources of information were themselves uncertain about the probabilities.

> All major media outlets (wire services, broadcast networks, news magazines, and metropolitan daily newspapers) [should] hire and train specialists who have more than a passing familiarity with reactors and the language of radiation. All other news media, regardless of their size, located near nuclear power plants should attempt to acquire similar knowledge or make plans to secure it during an emergency.

Today, many or most social phenomena are interrelated, to a greater or lesser degree, with science and technology. Consequently, any newspaper that really wants to provide accurate coverage of events must, in my opinion, employ journalists with a greater knowledge of science and technology than is commonly the case today. Japanese companies and national offices are sensitive to the need for scientific expertise. The science department of the *Asahi Shimbun*, for example, currently has a staff of seventeen reporters. All of them studied science and technology at universities. The size of this science staff is still small, however, when compared with the 100 or so reporters in the local news department, the 50 reporters in the political department, or the 45 reporters in the business and financial department. The *New York Times*, however, has only about seven reporters on its science desk. All are obviously excellent science reporters, but this small number cannot be expected to provide as much coverage as seventeen reporters. Still, the *Times* does a good job and its special science supplement has done much for education.

At the *Asahi Shimbun*, again, the newspaper's president, Seiki Watanabe, graduated from the agriculture department of his university, and was assigned to nuclear power and science and technology when he worked on the editorial staff of the paper. Moreover, the former director of the paper's editorial board, Junnosuke Kishida, majored in aeronautical engineering in his university days. I suspect that few, if any, of the world's other major mass-circulation newspapers have top executives with similar backgrounds in the physical sciences. It would appear that Japanese newspapers are far better prepared than their American counterparts to cope with the present age of science and technology. The role played by the *Asahi Shimbun* in curing Japan's anti-science disease was of considerable importance — of that there can be no doubt. Scientifically trained people in the media and in most public organizations are essential in shaping mass attitudes toward science education and technology.

Grassroot Science Support

About six weeks after the launching of the space shuttle *Columbia*, a "Conference on Space Manufacturing" was held at Princeton University. Its sponsors included Dr. Gerald K. O'Neill, who advocates the colonization of space by mankind. Although the conference registration fee was $125, around 150 people attended. Discussions lasted until late at night for four days. According to O'Neill, the remarkable success of the space shuttle brought closer the realization of the "space factory" concept that he and others had been considering for many years. There was talk of plans to build robot-manned factories on the surface of the moon; of a project to place numerous factory satellites in earth orbit, sending their products back to earth; and of such future technological projects as a new type of solar generating station satellite. O'Neill, in his latest book, *The Technology Edge* (1983), devoted his first chapter to the technical challenge of Japan and how the United Stated should compete once again.

A talk by Georgetown University's Charles Chafer, entitled "Space Policy: The Context of Legislation," impressed me at the conference: "If we are truly in earnest about making a reality of the 'space factory' concept, then we have no choice but to organize a citizens campaign in support of space development." Why is such a citizens campaign essential? Chafer went on to explain in some detail. First, he said, President Reagan has announced an economic policy of cutting the budget and lowering taxes. Under this kind of policy, even spending on social benefits will be rapidly reduced. Under such circumstances, it is highly unlikely that the space program will be able to obtain increased funding. If we simply lie down and wait, he said, the money for building space factories will never get into the budget. That was Chafer's analysis of the current state of affairs. What should be done? His answer was, "start a private level citizens campaign."

During the campus uprisings of the late 1960's, students demanded that they be allowed to participate in the governance of the universities. For example, they demanded a voice in the selection of the university president, or the right to be represented at faculty meetings. This was known as participatory democracy and now is called private sector initiative.

Initial attempts to gain direct organizational participation were the anti-pollution movement, the consumer movement, and the anti-science and anti-technology movements. These campaigns frequently proved effective. Politicians took up many of their causes, and the effects were felt in the political process. An increasing number of academics and other intellectuals began talking as if this "participation" were a new or splendid idea, an ideal expression of democracy. The mass media also encouraged the notion of direct participation by the citizen in politics. Participatory democracy, like any interest group activity, has the defect of giving rise to social inequities. This is because direct participation in politics tends to be limited to "political" issues: those persons having a strong desire to participate actively, as well as sufficient spare time to do so, become active. The people who came in person to the offices of the *Asahi Shimbun* to protest the paper's coverage of nuclear power were certainly persons of that description. Most were housewives, retired people, and students; the protestors gave the impression of having in their midst hardly any people holding regular jobs. Suppose that newspapers were to yield to pressure from such small, unrepresentative groups, and compromise the objectivity of their reporting. Or suppose that political leaders were to be influenced only by the views propagated by one or another segment of the public having both the time and desire to "participate." The results could well be unfairness and injustice — not the way good media firms operate.

The findings of public-opinion surveys indicate that the views espoused by activist groups of this kind are often unrepresentative of the public as a whole. Take the case of nuclear power, for example. Polls by the *Asahi Shimbun* have asked members of the Japanese public: "Do you favor, or oppose, the encouragement of nuclear power as an energy source for the future?" Those "favoring it" have ranged from 55 percent to more than 60 percent of the total. The number replying "opposed" has been only 20 to 25 percent. Even in a survey taken after the Three Mile Island accident, these percentages did not change. Thus, two to three times as many Japanese favor nuclear power as oppose it. Yet, citizens' groups attempting to act directly in the political process are almost entirely on the anti-nuclear side. Not once has a group of activists *favoring* nuclear power marched into the offices of the *Asahi Shimbun* to express its position. Thus, there is a strong likelihood that activist groups, by intervening directly in the process of policy formation, are distorting and weakening the basic democratic principle of majority decision.

A similar situation has existed with regard to the anti-science disease in America. Public-opinion surveys of American attitudes toward science and technology were taken in 1972 and 1974, when anti-science had already become quite severe. One of the questions asked was: "Do you feel that science and technology have changed life for the better, or for the worse?" Those replying "for the better" amounted to 70 percent of the total in 1972, and 75 percent in 1974. By contrast, those answering "for the worse" were only 8 percent in 1972 and 5 percent in 1974. In addition, people were questioned about their attitudes toward various occupations and professions, in an attempt to rate the comparative prestige of these jobs. Physicians were ranked first by the public; second were scientists, and third were engineers. Following them (in order) came ministers, architects, lawyers, bankers, accountants, and business-people. When members of the public were asked whether science and technology had done more good or more harm, 52 percent in 1972 and 57 percent in 1974 answered "more good than harm." Advances in medicine were the example most often cited of the good done by science, while environmental destruction was the most frequently mentioned example of harm. Thus, the American people as a whole do not appear to view science and technology with hostility.

However, if one were to judge from America's newspapers, magazines, and television, until very recently, one would think that all Americans were hostile toward science and technology. That is because the press and television have given

150

much more attention to the opinions of activists, intellectuals, and others opposed or ignorant of science, who are outspoken, articulate, and politically engaged. Nevertheless, the opinions held by intellectuals, scholars, and prominent cultural figures are by no means always valid ones. There is no assurance that the activist minority demanding participation is thinking in a scientific manner, with the livelihoods and welfare of the general public in mind.

Charles Chafer's recommendations to those who support science and technology are being gradually implemented in the United States. For example, a group led by Elizabeth M. Whelan, of the Harvard School of Public Health, established the American Council on Science and Health (ACSH) in 1978. The Council's annual report for 1980 contained this statement:

> Since 1969, when cyclamate, the artificial sweetener, was banned, a number of food additives, pesticides, drugs, and industrial chemicals have been prohibited or restricted. In many cases, these actions were based on controversial interpretations of limited experimental evidence. Frequently, the negative impact of regulation had not been adequately considered. ACSH was created in an effort to restore a sense of balance to the discussion of environmental health issues.

The primary goal of ACSH is to provide administrative officials, elected officials, and consumers with balanced information on important scientific issues. In order to do this, ACSH publishes articles, pamphlets, and newsletters, sending them to a large number and variety of individuals. Scientists belonging to ACSH appear before Congressional committees and at public discussions and meetings, to provide accurate information.

The members of ACSH's steering committee and scientific advisory committee included a variety of experts from universities and research centers all over the country: nutritional scientists, physicians, toxicologists, agricultural scientists, pharmacologists, zoologists, food technologists, botanists, and economists. One of these scientists, Dr. Norman Borlaug, a leader of the so-called "green revolution," won the Nobel Prize for Peace in 1970. One of the issues taken up by ACSH was hyperactivity in children. Between 5 and 10 percent of American children suffer from a chronic inability to remain quiet, to stay in their seats in a classroom, or to concentrate on a given task. A California physician, Dr. Ben Feingold, declared in 1973 that hyperactivity is caused by salicylates and by artificial colors and flavors that are added to foods. Salicylates are substances that occur naturally in several fruits and vegetables, but Feingold contended that if foodstuffs containing salicylates, or artificial coloring or flavoring, were removed from their diets, hyperactive children would regain self-control and attentiveness. Large numbers of parents became convinced by this theory and tried it with their children, and their reports of success in curing hyperactivity came pouring in. However, when ACSH looked closely at these results, it discovered that this "food therapy" may have had a psychological effect upon the children involved. Salicylates and artificial colors and flavors have no direct relationship to excessive impulsiveness and activity in children, and the Feingold theory is not valid. ACSH issued these findings in an article that was sent to the press and television stations.

In another case, ACSH tried to inform the broadest possible public of the fact that the government's ban on the use of saccharine was unwise. The ban itself was one result of the so-called "Delaney clause" in America's Pure Food and Drug Act. Under this clause, which was sponsored by Congressman James J. Delaney and attached to legislation enacted in 1958, any substance that is found by experimentation to be capable of causing cancer in animals or human beings may not be licensed for use as a food additive. The uproar in 1969 over the artificial sweetener cyclamate also grew out of the Delaney clause. The source of that furor was a laboratory test showing that if large doses of cyclamate were given to mice, the animals developed tumors. However, for a human being to receive a proportionately equivalent quantity of cyclamate, he or she would have to drink 350 bottles of fruit juice containing cyclamate every day in order to receive a cancer-causing dose of the chemical.

Similar experiments were performed with saccharine as well. Again, a human being would have to drink 800 bottles a day of soft drinks containing saccharine in order to receive a dosage equivalent to the amount that was given to the laboratory mice. The human intake of soft drinks would have to begin at birth, and then only at age 70 or 80 would the human subject be likely to develop cancer. Under the working of the Delaney clause, however, the experiments with saccharine had proven that the chemical was a carcinogen, and so the federal government decided to prohibit its use as a food additive. Elizabeth Whelan and the other members of ACSH strove to inform the general public about the scientific issues raised by the application of the Delaney clause in the saccharine case. Just as the anti-science activists are constantly "using" the press and television to propagate their views and increase their influence, ACSH similarly emphasizes the importance of activities involving the mass media. To summarize research findings, ACSH will hold a press conference, and keep a steady stream of press releases flowing to newspapers and wire services. According to ACSH's annual reports, such major newspapers as the *Washington Post*, the *Wall Street Journal*, the *San Francisco Chronicle*, and the *New York Times* have all published articles and news stories based on material sent to them by ACSH. Furthermore, scientists belonging to ACSH have made every possible effort to appear on television and familiarize audiences with the scientific approach to issues.

The main office of ACSH is in New York City, but to enable the organization to keep in close touch with Congress, it also maintains an office in Washington, D.C. Using this office as a base, members and officials of ACSH meet at least once each month with legislators and Capitol Hill staff members who are concerned with science or food, in order to discuss any issues that may be current or pending.

Elizabeth Whelan herself visited Japan in March, 1980, in an effort to introduce her pro-science movement. She met with a large number of scientists, including specialists in such fields as food technology and nutrition, but the results were not satisfactory. Apparently most of the Japanese university scientists she met responded something like this: "If we were to speak or act in a way that appears to be an act of defiance toward the consumer movement, that would be a very serious matter in Japan. We would be denounced as tools of the food manufacturers or as mouthpieces for the government. Our students would boycott our lectures, and groups of activists would even come to our classes and demonstrate in protest. The only scientists in Japan who could organize a movement like ACSH would be the ones who have already retired from university teaching." Under the Reagan administration local and state groups have had to organize to protect their interests as federal funds and enforcement have relaxed or dried up. This is an extension of these grassroots trends. Unfortunately, without federal support or guidance (not politicization), the big changes needed might not take place fast enough.

In the state of Missouri, a Food and Energy Council has been created. San Francisco has a group called Citizens for Adequate Energy. In Illinois, there is the Energy Education Program. In California, People for Energy Progress has even organized marches in support of nuclear power, with placards and banners bearing such slogans as "Nation Needs Nuclear," "Nuclear Power, Safe and Economical," and "We Want Nuclear Power." At the same time, several organizations have been formed to actively promote the cause of pure science. For instance, Dr. Carl Sagan, the author of *Cosmos*, established The Planetary Society in the spring of 1981. This is a citizens group dedicated to surveying Jupiter, Saturn, and other planets of the solar system, as well as planets revolving around other stars in our galaxy. According to Sagan, the pace of America's planetary exploration efforts has slackened ominously. After the *Voyager 2* space probe's encounter with Saturn in August 1981, there will be a period of more than four years during which not a single photographic image will be sent back from the planets by any American spacecraft. Meanwhile, the Soviet Union also appears to be slowing the pace of its own planetary exploration program.

Even a fairly large-scale planetary survey program like the *Voyager* project would cost a mere one-tenth of one percent of the federal budget each year. Nevertheless, says Sagan, politicians claim that this cost is too high, and that the general public does not support space programs. Elected officials, therefore, are reluctant to budget funds for such purposes. Sagan himself recounts that when he met with one member of Congress and urged him to support funding for space exploration, the Congressman remarked that the only letters he had received in support of the Galileo project to explore Jupiter were sent by people too young to vote.

Sagan believes the public does support explorations of space. After all, his own book *Cosmos* was a phenomenal best seller both in the United States and in Japan. Basically, the American public has a deep and continuing fascination with science and technology. Evidence for this is the success of the science magazine *Science '80*, established in 1980 by the American Association for the Advancement of Science, and aimed at the general public: its circulation passed 600,000 copies in April, 1981. Another popular magazine called *Omni*, founded in 1971, publishes both science articles and science fiction; it had already achieved a circulation of 858,000 in 1981. A Japanese version is also published. Another science magazine, *Discover*, published by Time Incorporated, has a circulation of 400,000 to 600,000, while *Science Digest* sells more than 300,000 copies of each issue.

Despite the growing public interest in science and technology indicated by these numbers, American political figures still insist that "the voters won't support science and technology" or that "the space program isn't popular with the public." Perhaps the voices of the people are drowned out by the loud and constant cries of the pundits and intellectuals who are chronic sufferers from the anti-science disease and the "anti-space" sickness. The 1983 movie "The Right Stuff" didn't become the best-seller it was expected to be. It was in order to counter this lopsided dominance by the anti-science and anti-space minority that Carl Sagan founded The Planetary Society. His aim is to gather as many members from the general public as possible, and thus to demonstrate once and for all to official Washington that the great majority of the American people do indeed understand the importance of space exploration. As of

1984, however, all we have is a high distrust in Reagan's "star wars" and continued decline in science education.

Another pro-science group is one devoted to promoting macro-engineering. It was founded in the belief that E.F. Schumacher's theory, "small is beautiful," is exerting a harmful influence upon public policy and private thinking and should be countered by promoting the reinstatement of large-scale technology. The group was formed as the result of an appeal by Professor Frank P. Davidson of M.I.T. In technology, according to Davidson, largeness of scale results in far greater efficiency, and economizes on natural resources in many cases. Taking magnetic levitation vehicles as an example, Davidson declares:

> [America] has fallen way behind the Japanese and Europeans in its ability to carry out large projects. Let me give an example. Here at M.I.T. is the Francis Bitter National Magnet Lab, where magnetic levitation was developed. Trains equipped with it could fly thousands of miles per hour. They even built a model of one of these railroads seven years ago. The U.S. government said, "Very good, boys, you were a complete success," and moved on to other projects.

> But the Japanese and the Germans took the same research and applied it to their needs. Late in 1979 the Japanese national railroad announced that it had run a full-size mag-lev car at five hundred three kilometers per hour, a new world record. They plan to install it on their Tokyo-to-Osaka line. By 1990 you'll be able to make that four hundred-eighty-kilometer trip in one hour; even their Bullet Train now takes three and a half hours. Meanwhile look at what the United States is talking about for its Northeast Corridor. The most daring proposal is something the Japanese actually built fifteen years ago.

Davidson and his colleagues have held a conference annually since 1978, compiling each year's proceedings in book form. The titles of these annual volumes include *Macro-Engineering and the Infrastructure of Tomorrow*, and *How Big and Still Beautiful? Macro-Engineering Revisited*. These books contain a proposal for a subsurface railway network linking all the continents on earth; a description of a robot able to perform work in space; plans for a large metropolis in Africa; and a design for a satellite that would generate electricity from solar energy for use on earth. This group believes in the notion that "Big is Powerful," and is trying to persuade the public and the government to think in those terms.

It is my hope that more and more efforts like those I have described above will emerge in America, and that their effects will multiply. I hope that America can recover as quickly and completely as possible from the anti-science disease. I have repeatedly noted how, since the second World War, Japan has taken America as her model in so many respects and she is readily influenced by America even today. We have prospered together and must do so in the future.

SOURCES

By Chapter in Order of Citation

CHAPTER I

Description of the press corps' reaction to *Apollo 11* launching: "Buji o inoru chikyū o ato ni," *Asahi Shimbun*, July 17, 1969, p. 15.

President Reagan's message: NASA press release for the *Columbia* launching.

Simon Ramo, *America's Technology Slip* (New York: John Wiley and Sons, 1980), pp. 8, 42, 56.

John Noble Wilford, "Space and the American Dream," *New York Times Magazine*, April 5, 1981, p. 54.

Lionel H. Olmer's comments: "U.S. Trade with Japan," *New York Times*, June 23, 1981, p. D2.

Figure 1: Kōzai Kurabu (Rolled Steel Club), ed., *Tekkō no jissai chishiki* (Tokyo: Toyo Keizai Shimpō–sha, 1980), p. 165.

Figure 2: Kōzai Kurabu, *ibid.*, p. 164.

Figure 3: Japanese Ministry of International Trade and Industry (MITI), *Nippon no tsūshō hakusho 1980* (Tokyo: Tsūshōsangyō–shō, 1981).

Regarding industrial robots: Yukiko Okuma, "Head Start in Robot Development," *Asahi Evening News* (Tokyo), March 18, 1981, p. 3.

Figure 4: Hisao Kanamori, "Kyōkyū jūshi keizai-gaku o ōen suru," *Shūkan Tōyō Keizai*, June 6, 1981, p. 50.

Figure 5: Data from Nippon Jidōsha Kōgyō–kai (Japan Automobile Manufacturers' Association), 1981.

Information on the decline of America's automobile industry: William Norhaus, "Detroit: How the Mighty Have Fallen," *New York Times*, April 19, 1981.

Article on W. Edwards Deming: Steve Lohr, "He Taught the Japanese," *New York Times*, May 10, 1981.

Buyers' experience with poor-quality American automobiles and public opinion poll on automobiles: John Holusha, "Detroit's New Stress on Quality," *New York Times*, April 30, 1981.

Accident at the Takahama nuclear power plant: "Buhin misu datta Takahama 2-gō–ki," *Asahi Shimbun*, November 6, 1979, p. 3.

Advertisements for Japanese automobiles: John Holusha, "Japan Puts Dazzle in Its Car Ads," *New York Times*, June 24, 1981.

Laid-off workers in Detroit: Ivor Peterson, "Detroit Despairs of Regaining Jobs," *New York Times*, April 27, 1981.

Export controls on Japanese automobiles: Clyde H. Farnsworth, "Tokyo's Car Curbs Hailed in U.S., But Japanese Makers are Angered," *New York Times*, May 2, 1981.

Alvin Toffler, *The Third Wave* (New York: Bantam Books, 1981), pp. 14, 138-49.

Attitude of American companies toward the space shuttle: John Noble Wilford, "Industrialization of Space: Why Business is Wary," *New York Times*, March 22, 1981.

Market share of Japanese semiconductors: Andrew Pollack, "Federal Aid Sought for Semiconductors," *New York Times*, May 19, 1981.

Figure 6: Data from U.S. Department of Commerce as reported in the *New York Times*, May 19, 1981.

Explanation given by American semiconductor manufacturers: Andrew Pollack, "Singing the Semiconductor Blues," *New York Times*, May 24, 1981.

American semiconductor companies' reponse to recession: Thomas J. Lueck, "Makers of Materials for Semiconductors Hurt by Slow Sales," *New York Times*, June 3, 1981.

Tariffs on semiconductors: Clyde H. Farnsworth, "U.S. and Japan Plan Cuts in Semiconductor Tariffs," *New York Times*, May 12, 1981.

Nippon Electric Company's operations in America: "NEC Plans $100 Million U.S. Plant," *New York Times*, June 27, 1981.

Christopher Byron, "How Japan Does It," *Time*, March 30, 1981, pp.54-63.

William G. Ouchi, *Theory Z: How American Business Can Meet the Japanese Challenge* (New York: Addison-Wesley Publishing Company, 1981).

Comments on Ouchi's book: Steve Lohr, "Japan Business Books Selling," *New York Times*, May 2, 1981.

Richard T. Pascal and Anthony G. Athos, *The Art of Japanese Management: Applications for American Executives* (New York: Simon and Schuster, 1981).

Use of the term "samurai management": Robert B. Reich, "The Profession of Management," *The New Republic*, June 27, 1981, pp. 27-32.

CHAPTER II

Report to the President: President's Task Force on Science Policy, *Science and Technology: Tools for Progress* (Washington, D.C.: Government Printing Office, 1970), pp. 8-9.

Thales of Miletus, Hippocrates of Cos, Theodorus of Samos: Benjamin Farrington, *Greek Science* (London: Penguin Books, 1961), pp. 36, 37, 81.

Archimedes and Antoine Lavoisier: Isaac Asimov, *Breakthroughs in Science* (Boston: Houghton Mifflin Co., 1960), pp. 1-8, 58-66.

Luddite movement and other workers' protests: Malcolm I. Thomas, *The Luddites* (London: David and Charles, 1970).

Anti-science movement in Germany after World War I: Paul Forman, *Weimar Culture, Causality, and Quantum Theory, 1918-1927: Adaptation by German Physicists and Mathematicians to a Hostile Intellectual Environment*, Historical Studies in Physical Sciences, Vol III (Philadelphia: University of Pennsylvania Press, 1971).

Daniel J. Kelvis, *The Physicists* (New York: Alfred A. Knopf, 1978), pp. 178-84, 237-39.

Siegfried Giedion, *Mechanization Takes Command* (Oxford: Oxford University Press, 1948), pp. 714-15.

Jacques Ellul, *The Technological Society* (Vintage Books, 1964), pp. 78, 79, 92.

Information of Shoichi Yokoi: "Guamu-tō kiseki no seizon," *Asahi Shimbun*, January 26, 1972, p. 1.

Figure 7: Marc Porat, *The Information Economy: Definition and Measurement*, Office of Telecommunications Special Publication 77-12, (Washington, D.C.: Government Printing Office, 1977).

Colin Norman's attitude: "The New Industrial Revolution: How Microelectronics May Change the Workplace," *The Futurist*, February, 1981, pp. 30-42.

Japanese industrial robots: "Robbotto ōi ga shitsugyō dasanu Nippon," *Asahi Shimbun*, June 2, 1981 (evening edition), p. 2.

Theodore Roszak, "The Monster and the Titan: Science, Knowledge, and Gnosis," *Daedalus*, Summer, 1974, pp. 17-32.

Francis Bacon's anecdote of the temple: *The New Organon and Related Writings* (Library of Liberal Arts, 1960), pp. 50-51.

Theodore Roszak's criticisms of science: *The Making of a Counter Culture* (New York: Doubleday and Co., 1969).

Atsuhiro Shibatani, *Han-kagaku-ron* (Tokyo: Misuzu shobō, 1973), pp. 89.

Figure 8: Japanese Science and Technology Agency, *Showa 52-nendo kagaku gijutsu hakusho* (Tokyo: Science and Technology Agency, 1978), p. 3.

Illich's theory about bicycle transportation: Illich Forum, ed., *Jinrui no kibō* (Tokyo: Shin-hyōron, 1981), pp. 45, 53; and Ivan Illich, *Energy and Equity* (New York: Harper & Row, 1974), pp. 59-64.

CHAPTER III

"Carousel of Progress": Shigeru Kimura, *Atomu kisha sekai dōchū–ki* (Tokyo: Asahi Shimbun-sha, 1965), pp. 236-38, 243-49; and dialogue script provided by General Electric Corporation.

Soviet atomic bomb: Harry S Truman, "Statement by the President on Announcing the First Atomic Explosion in the U.S.S.R., September 23, 1949," and "President's News Conference of October 6, 1949," *Public Papers of the Presidents*, Harry S Truman, 1949 (Washington, D.C.: Government Printing Office, 1964), pp. 485, 503.

American, British, and Soviet atomic hydrogen bomb tests: *New York Times*, 1950-53.

Fukuryu Maru No. 5 incident: "Usu-kimi-warui sankaku kumo," *Asahi Shimbun*, March 17, 1954.

Figure 9: *Asahi Shimbun*.

Visits to laboratories: Kelvis, *The Physicists*, p. 396.

Dwight D. Eisenhower, "Farewell Radio and Television Address to the American People, January 17, 1961," *Public Papers of the Presidents*, Dwight D. Eisenhower, 1960-61 (Washington, D.C.: Government Printing Office, 1961), pp. 1038-39.

Meeting between Eisenhower and Kistiakowski: George B. Kistiakowski, *A Scientist at the White House* (Cambridge, Mass.: Harvard University Press, 1976), pp. 424-25.

Rachel Carson's life: Frank Graham, Jr., *Since Silent Spring* (Boston: Houghton Mifflin, 1970), pp. 4-20.

Quotation from Rachel Carson, *Silent Spring* (Boston: Houghton Mifflin, 1962), pp. 2-3.

Sales history of *Silent Spring*: Graham, *Since Silent Spring*, p. 69.

Legislation on pesticides: Graham, *Since Silent Spring*, p. 72.

Comments by Eric Sevareid: Graham, *Since Silent Spring*, p. 79.

Thalidomide incident: Robert K. Plumb, "Deformed Babies Traced to Drug," *New York Times*, April 12, 1962, p. 37; and "Sleeping Pill Nightmare," *Time*, February 23, 1962, p. 86.

Rachel Carson's comments on herbicides: *Silent Spring*, pp. 34-5.

America's defoliation campaign in Vietnam: "Kareha sakusen kagaku hōkoku," *Shizen*, February, 1970, pp. 49-76.

Open letter to President Johnson: "A Letter to the President of the United States" (advertisement), *New York Times*, March 31, 1965, p. A18.

Ralph Nader's career and publications: Charles McCarry, *Citizen Nader* (New York: Saturday Review Press, 1972), pp. 3-21.

Nader's criticisms of Detroit's engineers: Ralph Nader, *Unsafe at Any Speed, Updated* (New York: Grossman, 1972), p. 28.

Critical comments about the space program: Shigeru Kimura, *Tsuki ni idomu jikken-shitsu* (Tokyo: Asahi Shimbun-sha, 1966), p. 240.

Editorial by Philip H. Abelson, "National Science Policy," *Science*, January 28, 1966, p. 407.

John F. Kennedy's message to Congress: "Special Message to the Congress on Urgent National Needs, May 25, 1961," *Public Papers of the Presidents*, John F. Kennedy, 1961 (Washington, D.C.: Government Printing Office, 1962), p. 669.

Kennedy's address at Rice University: "Address at Rice University in Houston on the Nation's Space Effort, September 12, 1962," *Public Papers of the Presidents*, John F. Kennedy, 1961 (Washington, D.C,: Government Printing Office, 1962), p. 700.

Comments by Sen'ichiro Hakomori: "Kaigai kara mita Nippon no seikagaku," *Seikagaku*, February, 1981, pp. 68-9.

Lyndon B. Johnson, "Statement by the President Following a Meeting to Review Goals for Medical Research and Health Services, June 27, 1966," *Public Papers of the Presidents*, Lyndon B. Johnson, 1966 (Washington, D.C.: Government Printing Office, 1967), p. 652.

Comment by John Fischer, "Why Our Scientists Are About to be Dragged, Moaning, Into Politics," *Harper's*, September, 1966, p. 16.

Project Mohole: "Project Nohole?" *Newsweek*, August 29, 1966, pp. 60-1; and "No Mohole?" *Scientific American*, July 1966, pp. 48-9.

Editorial by Philip H. Abelson, "Penny Wise, Pound Foolish," *Science*, June 3, 1966.
Data on printing history of the Japanese edition of *Silent Spring*: courtesy of the publisher, Shinchō–sha, Tokyo. I attempted to obtain exact figures of the latter from Carson's publisher, Houghton Mifflin, but they declined to release the information on the grounds that it is a trade secret.
Japanese attitudes toward nature: Japanese Ministry of Education, Institute of Statistical Mathematics, *Kokumin-sei no kenkyū: dai-6-kai zenkoku chōsa* (Tokyo: Ministry of Education, 1979), p. 26.

CHAPTER IV

Free speech movement: articles in the *New York Times*.
Analysis of campus unrest: Seymour M. Lipset, *Rebellion in the University* (Chicago: University of Chicago Press, 1976), pp. viii-xxvi.
Anti-science tone of the leadership of the campus unrest: Takichi Shimizu, ed., *Han-taisei no shisō* (Tokyo: Jiyū Kokumin-sha, 1970), pp. 219-20.
Criticism of Marx for being pro-science: Takichi Shimizu, *ibid.*, pp. 93-9.
Remarks by the crew of *Apollo 8*: "Excerpts from Radio Conversations between the Apollo 8 Crew and Houston," *New York Times*, December 5, 1968.
Black demonstrators at Cape Canaveral and Houston: "Kokujin ga kōgi no demo," *Asahi Shimbun*, July 16, 1969 (evening edition), p. 4.
Lewis Mumford, "No: A Symbolic Act of War," *New York Times*, July 21, 1969.
Comments by Taruho Inagaki: *Asahi Shimbun*, July 21, 1969 (evening edition), p. 11.
Comments by Aiko Sato: *Asahi Shimbun*, July 25, 1969, p. 4.
Richard Nixon's statements on environmental issues, "Remarks on Signing the National Environmental Policy Act of 1969" and "Statement about the National Environment Policy Act of 1969," *Public Papers of the Presidents*, Richard Nixon, 1970 (Washington, D.C.: Government Printing Office, 1971), pp. 1-3.
Presidential Task Force report on the environment: *Report of the Task Force on Resources and Environment*, December 5, 1968 (unpublished document), p. 2.
Earth Day observances: "Millions Join Earth Day Observances Across the Nation," *New York Times*, April 23, 1970.
Editorial on Earth Day: "The Good Earth," *New York Times*, April 23, 1970, p. 36.
Nixon's attack on technology, "Message to the Congress Transmitting the First Annual Report of the Council on Environmental Quality, August 10, 1970," *Public Papers of the Presidents*, Richard Nixon, 1970 (Washington, D.C.: Government Printing Ofice, 1971), p. 654.
Article on defective foreign automobiles: John D. Morris, "Publicity Avoided on Some Recalls," *New York Times*, May 12, 1969, p. 34.
Japanese press coverage of defects in Japanese cars: "Kekkan naze kakusu, Nippon no jidōsha," *Asahi Shimbun*, June 1, 1969, p. 15.
Ban on cyclamate: Harold M. Schmeck, Jr., "Government Officially Announces Cyclamate Sweeteners will be Taken Off Market Early Next Year," *New York Times*, October 19, 1969, p. 58.

Japanese reaction to the cyclamate ban: "Jinkō kanmiryō chikuro, Bei de zenmen kinshi," *Asahi Shimbun*, October 19, 1969.

Article by Hitoshi Aiba: *Asahi Shimbun*, October 2, 1971, p. 12.

Newspaper publishers' convention: Thomas F. Brady, "Publishers Rate Environment as the 'Big Story' in U.S.," *New York Times*, April 23, 1970.

Quotation from Paul R. Ehrlich's Prologue: *The Population Bomb* (New York: Ballantine Books, 1971), p. xi.

Ehrlich's criticism of science and technology: *The Population Bomb*, p. 156.

Quotations on technology from Donella H. Meadows, ed., *The Limits to Growth*, 2nd. ed. (New York: Universe Books, 1974), pp. 145, 154.

Club of Rome's comments on exhaustion of resources and Commentary by the Club of Rome: Meadows, *The Limits to Growth*, pp. 56-58, 187.

Statement by Bentley Glass: Walter Sullivan, "Academic Meetings: A Scientist Finds Science Limited," *New York Times*, December 29, 1970.

Shibatani, *Han-kagaku-ron*, p. 105, 107.

Statement by Michelson: Morris Gorran, *Science and Anti-Science* (Ann Arbor: Ann Arbor Science, 1974), p. 115.

Views of Lord Kelvin: Harvey Brooks, "Can Science Survive in the Modern Age?" *Science*, October 1, 1971, p. 25.

CHAPTER V

History of the SST Project: Mel Horwitch, "The American SST: A Cautionary Analysis," *Macro-Engineering and the Infrastructure of Tomorrow* (Boulder, Colo.: Westview Press, 1978), pp. 139-76; Joel Primack and Frank von Hippel, "Scientists, Politics, and the SST: A Critical Review," *Bulletin of the Atomic Scientists*, April, 1972, pp. 24-30; John M. Logsdon, "Contemporary Anti-Technology Sentiments: The Case of the SST," *Annual Meeting of the Society for the History of Technology*, 1972, pp. 17-31.

John F. Kennedy's SST Statement: "Remarks at Colorado Springs to the Graduating Class of the U.S. Air Force Academy, June 5, 1963," *Public Papers of the Presidents*, John F. Kennedy, 1963 (Washington, D.C.: Government Printing Office, 1964), pp. 140-41.

Maurice H. Stans: "Wait a Minute," *U.S. Department of Commerce News*, July 15, 1971.

Stans' memorandum to Nixon: "Costs of Environmental Compliance" (Memorandum for the President, June 18, 1971).

Russell E. Train's memorandum to Nixon: "Environmental Quality and Economic Progress" (Memorandum for the President, June 30, 1971).

Paul W. McCracken's memorandum to Nixon: (Memorandum for the President, July 20, 1971).

Article attacking the Bodega Bay nuclear plant: Gene Marine, "Outrage on Bodega Head," *The Nation*, June 22, 1963, pp. 524-27.

Editorial by Philip H. Abelson, "Nuclear Power — Rosy Optimism and Harsh Reality," *Science*, July 12, 1968, p. 113.

Figure 13: Nippon Genshiryoku Sangyō Kaigi (Japan Atomic Industrial Forum, Inc.), ed., *Genshiryoku hatsudensho ichiran-hyō* (Tokyo: JAIF, 1981), p. 24.

Ernest J. Sternglass article on fallout: "Infant Mortality and Nuclear Tests," *Bulletin of the Atomic Scientists*, April, 1969, pp. 18-20 and "The Reply," *Bulletin of the Atomic Scientists*, October, 1969, p. 32.

Article on nuclear generating plants: "The Dilemmas of Power," *Time*, August 29, 1969, pp. 38,39.

John W. Goffman and Arthur R. Tamplin, "Radiation: The Invisible Casualties," *Environment*, April. 1970, pp. 12-49.

Nader's opposition to nuclear power: Hays Gorey, *Nader and the Power of Everyman* (New York: Grosset and Dunlap, 1975), p. 96.

Pro-nuclear article by a Japanese socialist: Shigeyoshi Matsumae, "Genshiryoku heiwa riyō no hōkō," *Gekkan Shakai-tō*, April 1957, pp. 74-80.

Newspaper article on Arthur Tamplin: "Stoppu genpatsu no itsuka-kan, Tanpurin hakase, jūmin undō ni 'katsu'," *Mainichi Shimbun*, July 30, 1973.

Mutsu affair: "Mutsu," 60-part series in *Asahi Shimbun*, March 1, 1977, *et seq.*

E.F. Schumacher's views on technology amd nuclear power: *Small is Beautiful* (New York: Harper Colophon Books, 1973), pp. 31, 32, 127, 135, 150.

Critique of Schumacher's views: Shigeru Kimura, "Kyodai gijutsu no fukken," *Tokyo Review*, 69 (1980).

Amory E. Lovins' article: "Energy Strategy: The Road not Taken," *Foreign Affairs*, October 1976.

Amory E. Lovins, *Soft Energy Paths: Toward a Durable Peace* (New York: Friends of the Earth, 1977).

Critique of Lovins' views: Shigeru Kimura, "Today's Science: 'Soft energy path' just a daydream", *Asahi Evening News*, April 15, 1980.

Comments by Saburo Okita: Foreword to Amory E. Lovins, *Sofuto enerugii pasu* (Tokyo: Daiyamondo-sha, 1980).

Opposition to petro-proteins and to lysine: articles in the *Asahi Shimbun.*

Comments by the director of the National Genetics Institute: Yatarō Tajima, Kankyō wa iden ni dō eikyō suru ka (Tokyo: Daiyamondo-sha, 1981), pp. 181-82.

CHAPTER VI

Unsuccessful attempt to assassinate President Reagan: reports in the *New York Times*, 1981.

Memorandum on robbery: Memorandum sent to the author and other Fellows at the Woodrow Wilson International Center for Scholars, Washington, D.C., dated March 4, 1981.

Astrology column on wounding of President Reagan: Svetlana Godillo, "The Aspects of Reagan's Chart," *Washington Post*, April 5, 1981.

History of astrology and comments on astrology columns: Carl Sagan, *Cosmos* (New York: Random House, 1980), pp. 48-55.

James A. Michener's comments on astrology columns and on the *I Ching*: "Comments on the Anti-Science Epidemic," *Social Education*, May 1980, pp. 376, 378.

162

Figure 14: Number of articles on UFOs based on the annual volumes of the *New York Times Index*.

Accounts of UFO sightings: National Military Establishment, Office of Public Information, *Project Saucer*, Memorandum to the press, dated April 27, 1949.

UFO outbreak in 1966: reports in the *New York Times*, 1966.

University of Colorado UFO study: Walter Sullivan, "Scientists Back Report on
• UFOs," *New York Times*, January 9, 1969, p. 36.

UFO sighting by Jimmy Carter: "Notes on People," *New York Times*, September 14, 1973, p. 35.

Carter's request that NASA study UFOs: "Carter Asks Space Agency to Investigate UFOs," *New York Times*, November 27, 1977, p. 26.

NASA's refusal: "NASA Refuses to Reopen Investigation of UFOs," *New York Times*, December 28, 1977, p. 14.

Sagan's views on UFOs: *Cosmos*, p. 292.

Loch Ness monster: Shigeru Kimura, "Nesshii wa honto ni iru ka?" in *Asahi shōnen shōjo rika nenkan* (Tokyo: Asahi Shimbun-sha, 1977), pp. 348-55; Roy P. Mackal, *The Monsters of Loch Ness* (London: McDonald and Jane's, 1976).

Rine's photograph: accounts in the *New York Times*, December 5, 1975.

Scientific christening of "Nessie": "Naming the Loch Ness Monster," *Nature*, December 11, 1975, pp. 466-68.

Muir's evaluation of Rines's photograph: "Loch Ness Expert Thinks Photos Show Movie Dummy," *New York Times*, December 14, 1975, p. 10.

Career of Uri Geller: "Geller, Uri," *Current Biography, 1978* (New York: Wilson, 1978), pp. 152-55.

Uri Geller, *My Story* (New York: Praeger, 1975).

"Paranormal" children in Japan: Shigeru Kimura, "Bakageta supuun-mage sōdō" in *Asahi shōnen shōjo rika nenkan* (Tokyo: Asahi Shimbun-sha, 1975), pp. 382-85.

Harold Puthoff and Russell Targ, "Information Transmission under Conditions of Sensory Shielding," *Nature*, October 18, 1974, pp. 602-7.

Joseph Hanlon, "Uri Geller and Science," *New Scientist*, October 17, 1974, pp. 170-78.

California evolution trial: Robert Rindsey, "Creationists Gather to Try Toppling Darwin's Pedestal," *New York Times*, March 1, 1981, p. E21.

Casey Segraves' testimony: Philip J. Hilts, "I Believe God Created Man," *Washington Post*, March 4, 1981.

Arkansas legislation on creationism: "Bill on Creationism Passes in Arkansas," *New York Times*, March 18, 1981, p. A16.

Isaac Asimov, "The 'Threat' of Creationism," *New York Times Magazine*, June 14, 1981, pp. 100-1.

Boom in religious books: Edwin McDowell, "Religious Publishing: Going Skyward," *New York Times*, May 12, 1981.

CHAPTER VII

Edward J. Burger, Jr., *Science at the White House: A Political Liability* (Baltimore: Johns Hopkins University Press, 1980), p. 2, 87, 105-13.

David Z. Beckler, "The Precarious Life of Science in the White House," *Daedalus*, Summer 1974, p. 115, 124, 127.

Luna 15 and *Luna 16*: Shigeru Kimura, *Uchū e no dōhyō* (revised ed.; Tokyo: Kyōritsu Shuppan, 1976), p. 134.

Interview with Braun: Shigeru Kimura, *Tsuki ni idomu jikken-shitsu* (Tokyo: Asahi Shimbun-sha, 1966), pp. 235-39.

Agnew's Mars statement: "Tsugi wa kasei mezasu," *Asahi Shimbun*, July 14, 1969, p. 1.

Braun's press conference: "Apollo 11 Postlaunch Press Conference" (NASA transcript, dated July 14, 1969), p. 11B/2.

Comments by Sony's Masaru Ibuka: "Kaihatsu mokuhyō–sadame mōzen to idomō," *Asahi Shimbun*, August 2, 1981, p. 8.

Figures 15, 17-21, 23: National Science Board, *Science Indicators, 1978* (Washington, D.C.: Government Printing Office, 1979), pp. 6, 7, 8, 17, 32, 44, 50.

Tokai-mura reprocessing facility: reports in the *Asahi Shimbun*.

Figures 16, 22: Japanese Science and Technology Agency, *Kagaku gijutsu hakusho 1979-nenban* (Tokyo: Science and Technology Agency, 1979), pp. 44, 161.

Moyers-Adams television interview: *Bill Moyers' Journal: Defense, Dissent, and the Dollar* (New York: Educational Broadcasting Corporation, 1981), pp. 5-6.

Comments by Blommer and Banner: Randolf E. Schmid, "U.S. Grip on New Technology Slipping to Foreign Inventors," *Asahi Evening News*, June 20, 1981, p. 4.

Table 2: Report on science education: National Science Foundation and U.S. Department of Education, *Science and Engineering Education for the 1980's and Beyond* (Washington, D.C.: Government Printing Office, 1980).

David Savage, "The Growing Science Gap in our Schools," *Washington Post*, May 31, 1981.

William G. Aldridge, "U.S. Secondary School Science Education: Approaching the Dark Age," *Fusion*, March-April, 1981, p. 8.

CHAPTER VIII

John Noble Wilford, "Space and the American Vision," *New York Times Magazine*, April 5, 1981, p. 53.

Analogy of the development of the American West: John Noble Wilford, "The Industrialization of Space: Why Business is Wary," *New York Times*, March 22, 1981.

William Stockton, "The Technology Race: America's Struggle to Stay Ahead," *New York Times Magazine*, June 28, 1981, pp. 14, 17.

Japanese reaction to the oil crisis: coverage in the *Asahi Shimbun*.

Quotation from Tahar's book: Sōichirō Tahara, *Seizon no keiyaku* (Tokyo: Bungei Shunjū, 1981), pp. 147-51.

Yukiko Okuma's "Afterword" to her book: *Kaku nenryō: tansa kara haikibutsu shori made* (Tokyo: Asahi Shimbun-sha, 1977), p. 305.

Irradiated potatoes: coverage in the *Asahi Shimbun*.

"New Nessie" sea monster: coverage in the *Asahi Shimbun*.

Quotation from the Presidential Commission's report: *Report of the President's Commission on the Accident at Three Mile Island* (Washington, D.C.: Government Printing Office, 1979), pp. 19, 58, 79.

Charles M. Chafer, "Space Policy: The Context of Legislation" (Paper delivered at the Princeton Conference on Space Manufacturing, May 1981).

Japanese opinion polls: *Asahi Shimbun*, January 3, 1981.

American Opinion polls: Harris Survey news release dated February 17, 1972; and National Science Board, *Science Indicators, 1974* (Washington, D.C.: Government Printing Office, 1975).

ACSH statement of purpose: American Council on Science and Health, *Second Annual Report* (New York: ACSH, 1980).

Hyperactive children: American Council on Science and Health, *Diet and Hyperactivity: Is there a Relationship?* (New York: ACSH, 1980).

Saccharine ban: Eiji Mikado "Hamon yobu Beikoku no sakkarin kinshi," *Asahi Shimbun*, April 3, 1977.

Carl Sagan, "The Adventure of the Planets," *Planetary Report* I, 1 (1981).

Popularity of science magazines: Philip H. Dougherty, "Science Magazines Taking Off," *New York Times*, April 2, 1981.

Davidson on magnetic levitation: Dan Ross, *et al.*, "Macro," *Omni*, April, 1981, p. 118.

Books on macro-technology: Frank P. Davidson, ed., *et al.*, *Macro-Engineering and the Infrastructure of Tomorrow* (Boulder, Colo: Westview Press, 1978) and *How Big and Still Beautiful? Macro-Engineering Revisited* (Boulder, Colo.: Westview Press, 1980).

Books Co-Published by
University Press of America and
The Woodrow Wilson International Center for Scholars

KOREAN STUDIES IN AMERICA: OPTIONS FOR THE FUTURE
Edited by Ronald A. Morse

**THE NATIONAL INTEREST OF THE UNITED STATES
IN FOREIGN POLICY**
Edited by Prosser Gifford

**REFLECTIONS ON A CENTURY OF
UNITED STATES-KOREAN RELATIONS**
Edited by Ronald A. Morse

**VIETNAM AS HISTORY: TEN YEARS AFTER
THE PARIS PEACE ACCORDS**
Edited by Peter Braestrup

SOUTHEAST ASIAN STUDIES: OPTIONS FOR THE FUTURE
Edited by Ronald A. Morse